Mathematics for the Imagination

Peter Higgins is Professor of Mathematics at the University of Essex. He is also the author of the successful *Mathematics for the Curious* (OUP, 1998), amongst other titles.

Mathematics for the  Imagination

Peter M. Higgins

OXFORD
UNIVERSITY PRESS

OXFORD

UNIVERSITY PRESS

Great Clarendon Street, Oxford OX2 6DP

Oxford University Press is a department of the University of Oxford.
It furthers the University's objective of excellence in research, scholarship,
and education by publishing worldwide in

Oxford New York

Auckland Bangkok Buenos Aires Cape Town Chennai
Dar es Salaam Delhi Hong Kong Istanbul Karachi Kolkata
Kuala Lumpur Madrid Melbourne Mexico City Mumbai Nairobi
São Paulo Shanghai Singapore Taipei Tokyo Toronto

with an associated company in Berlin

Oxford is a registered trade mark of Oxford University Press
in the UK and in certain other countries

Published in the United States
by Oxford University Press Inc., New York

British Library Cataloguing in Publication Data

Data available

Library of Congress Cataloging in Publication Data

Data available

ISBN 978-0-19-860460-0

4

Typeset in Minion 10.25/12pt
by Kolam Information Services Pvt. Ltd., Pondicherry, India

Printed in Great Britain
by Clays Ltd.,
Bungay, Suffolk

Contents

Preface

The main purpose of this book it to convey to my audience the history and development of mathematics throughout the ages and to explain some of its most interesting features. The emphasis is on the visual so there is little in the way of traditional formula and that which there is should not interrupt the flow of the material. The final chapter is exceptional and has been included on request so that the book can be of greater value for those making a more serious study of the subject who would like some detail without having to pursue some other source. I would like to thank the staff and readers of Oxford University Press together with my daughters Genevieve and Vanessa who read and commented on earlier drafts of *Mathematics for the Imagination*.

The text is written so that it can be read straight through but the reader should feel under no obligation to do that and may well enjoy dipping into it as they fancy. I hope that everyone can find something that captures their imagination somewhere it the text—if not then I really will have failed as mathematics is the deepest and most fascinating subject in the world.

Peter Higgins,
Colchester, 2002.

○ World Travel

The story of mathematics is a big story that began thousands of years ago when people first began to count their livestock or measure the extent of a field. Since that time the mathematical character of the world has revealed itself at every turn. For those who seek the truth, mathematics opens a doorway to understanding that is both deep and undeniable and those trying to comprehend the physical world cannot but be impressed by its effectiveness. All of us are naturally intrigued by anything genuinely interesting or surprising and mathematics will always score highly in that regard. At the same time however we can be intimidated by the uncompromising nature of the subject but, be assured, this is everyone's experience. One of the bonuses of living in our own age is that it is possible to understand matters that were a complete mystery to even the most learned and brilliant of our ancestors. The mind of a modern person can be much more powerful than the minds of past generations because we have at our disposal enlightening ideas and techniques that let us comprehend all manner of subjects, not only in broad principle but often in fine, practical detail. This is a rewarding privilege and one that is well worth making use of from time to time.

Not everyone likes working the soil but we can all enjoy the beauty of a splendid garden, and so it is also with a world of thought that comes about only after much toil on someone's part: once created, it can be delightful to be guided through to view its highlights, and witness its achievements. The world we shall be looking at in this book is one of mathematics, but you will not come across too much in the way of traditional mathematical formulae. All the same you will find plenty here to see and enjoy and we shall often make use of pictures: a simple picture, even one you draw yourself, can do wonders to clarify an otherwise fuzzy idea.

This is a journey of the imagination undertaken for its own sake rather than with the need to arrive at any special destination. Being more in the nature of a leisurely walk, it will have something of a meandering air, for there are no time limitations and no one is bound to follow the guide.

We sometimes imagine that we prefer concrete, real-world situations to abstract ones. Often however it is only that we prefer the familiar to the less familiar. There can hardly be anything more 'real-world' than the world itself so let us begin with the idea of the globe and travelling around it.

Imagine that you are flying directly from London to San Francisco. You might thumb through the airline magazine and discover a colourful map of the world. This offers a very distorted and abstract picture of the real world but it is one with which we all feel comfortable enough never the less. The magazine tells the passengers that their destination, San Francisco, is thousands of miles west of London and somewhat to the south as well. You are therefore entitled to feel a little bemused when your Captain announces half an hour into the journey that you are overflying Manchester. A couple of hours later you may be positively amazed while sipping your after-dinner drink as you look down to find that you seem to have

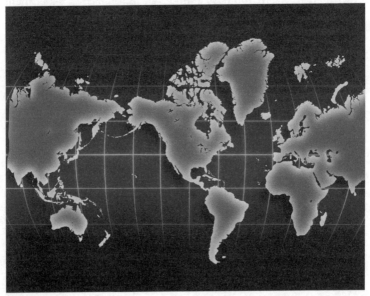

Fig. 1.1: Airline map of the world

arrived at the North Pole! The glaciers of Greenland may look spectacular but what on earth are we doing here? Our faith in aircraft and their crews is such that a seasoned traveller suffers no real alarm at finding himself miles above the arctic ice cap and may well nod off to sleep. But what if sleepy confusion is not good enough for us and we insist on an answer: what is going on? If we want to go west and south, why head north?

The answer lies in the fact that you are following the shortest path to your destination. We may all know that the shortest distance between two points *on a flat surface* is a straight line but our planet is far from flat for we live on a giant sphere. The straight line joining London to San Francisco goes underground through the Earth itself so is no path for an aeroplane. What then is the nature of the best line of flight?

A shortest path between two points on a surface is known as a *geodesic* and the geodesics of a sphere are sections, usually called *arcs*, of what are known as *great circles* of which lines of longitude form examples but, with the exception of the equator, the parallels of latitude do not. If you have a model globe to hand an experiment that demonstrates this is to hold a piece of elastic to the surface between the take-off point and the destination, London and San Francisco respectively in our case. The elastic will find this optimal great circle path of its own accord as, being the shortest, it is the path under which the elastic is least stretched.

The great circles on the surface of the globe all share a common centre, that point being the centre of the Earth itself. Each divides the sphere into two equal parts or *hemispheres*. On an ordinary stationary sphere, none of them has any natural priority over the rest. Our earth though has been spinning on a fixed axis since the day it was born billions of years ago. For that reason, the great circles of *longitude*, the ones that pass through the poles, are of special significance. The equatorial plane contains the *equator*, the one and only great circle at right angles to the axis of rotation. The plane of all other great circles makes some arbitrary angle with the equator. Between any two points on the surface of the Earth there is just one great circle that runs between them, that being formed by the meeting of the planet's surface with the plane that runs through the two points and the Earth's centre. The shortest distance between the two given points P and Q on the surface is the shorter of the two arcs of this great circle between the points: see Figure 1.2 where C represents the Earth's centre.

Imagine now, for example, that you are in London (at 0° longitude) and you wish to travel to the point, located in the North Pacific Ocean, at the same latitude but 180° east, that is to say you want to travel to the point on the same latitude line exactly opposite London on the globe.

The shortest path involves following the great circle of longitude straight over the North Pole and down again. In other words travelling first north and then south is the quickest way to move 180° east. Similarly the great circle between London and San Francisco, while not travelling over the pole, certainly crosses well inside the arctic circle. Examining a model globe will soon convince you of this, although studying your airline map might not. With a flat map the temptation is simply to draw a straight line between the two points of your flight which seems much more natural than to take the great circle path which would have you drawing a broken curve that could

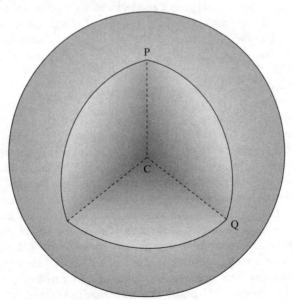

Fig. 1.2: Arc of a great circle on the globe

slip off the north edge of the map and re-enter the picture at another position along the northern boundary. Although you may *feel* happier looking at the flat map, its distortion of the real world is the source of confusion.

So we see that the shortest path from London to 180° east is *not* the one you would follow if you simply set your compass bearing due west and followed that line of latitude. The path that you take when you follow a fixed compass bearing is known by sailors as the *rhumb line* to your destination. Such a line, though it is very much a curved line, meets every *meridian* (the term for a line of longitude, that is a great semicircle from pole to pole) and every *parallel* (line of latitude) at a constant angle. For example, if you were to begin in Hawaii, and set your compass bearing north east, the rhumb line that you would travel would have you meet every meridian and latitude line you encounter equally at an angle of 45°. However, if you continue to follow this bearing you will never end up back where you began. This is because, although you can continue heading east as long as you wish, you cannot go north forever for eventually you must end up at the most northerly point on the globe—the North Pole. This shows you that a rhumb line is not the same thing as a great circle; it is not even a circle but rather a spiral. When projected on to a flat representation of the Earth viewed from above the North Pole the rhumb line gives a picture as shown in Figure 1.3.

The spiral pictured is known as an *equiangular spiral* as tangents to the curve make a constant angle with the meridian lines which project out from the pole to the outer circle representing the equator. The nature of this rather special curve, so important in navigation, was discovered by the English mathematician Thomas Harriot around the end of 16th century. Harriot seems to have been a very able scientist indeed but the true worth of his achievements did not become widely recognized, perhaps because his reputation suffered through involvement with intrigues surrounding his glittering friends, Sir Walter Ralegh (executed in 1618), the playwright Christopher Marlowe (apparently murdered in a tavern brawl in 1593), and Guy Fawkes (hung, drawn, and quartered in 1606). His only published work during his lifetime concerned Virginia where he visited and made a serious study of Native American languages and customs, although this early 1585 attempt at settlement, situated in

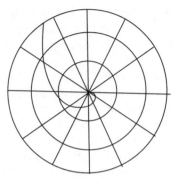

Fig. 1.3: Equiangular spiral: view of a rhumb line from above the North Pole. Longitude lines radiate from the pole while the concentric circles represent parallels of latitude

modern-day North Carolina, failed. Harriot's curve also has a stunning self-similarity property in that if you expand it by a fixed factor, for example if you take each point on the curve and double its distance from the pole while remaining on the same meridian, you get the same curve back again, only rotated through a certain angle that depends on the multiplier used. The celebrated Swiss, James Bernoulli, one member of mathematics' most illustrious family, was so taken by this that he arranged for his tombstone to be decorated by the spiral and accompanying epitaph which translates from the Latin as 'though changed, I arise again the same'.

Of the parallels of latitude, the equator is naturally pre-eminent, being the only one that is a great circle. In contrast, the meridians have equal status so to select one of them as the *prime meridian* is arbitrary. The establishment of London, to be precise Greenwich Observatory, as the line of 0° longitude has given London the permanent international status as the place where east meets west although every point along this meridian shares this status with Greenwich—you might have noticed that the spot on the earth's surface of 0° longitude and latitude lies in the equatorial Atlantic and is of no other significance.

The recognition of Greenwich as the home of the prime meridian evolved through Britain becoming a major seafaring nation in the 17th and 18th centuries and through the particular work of the Astronomer Royal, Nevil Maskelyne, who fixed Greenwich as the seat of the 0° longitude line and drew up detailed tables that were widely used by sailors throughout the world in the late 18th and early 19th centuries. It became common practice for the ship's captain to calculate his longitude with respect to Greenwich rather than some other alternative such as the starting point of his voyage. The International Meridian Conference of 1884 in the US capital Washington finally settled Greenwich as the prime meridian for all time, despite efforts by the French to have Paris Mean Time adopted as the international standard.

Greenwich Mean Time is the standard throughout the world although ironically, for more than half the year, Greenwich has its clocks adjusted to British Summer Time which is an hour ahead of GMT. Credit for the idea of daylight saving itself goes to the American scientist and co-founder of the United States Constitution, Benjamin Franklin.

For every 15° that you travel east the sun rises one hour earlier (as 15° × 24 = 360°). This has a number of intriguing consequences, amusing and otherwise. If you are very near the poles, the lines of longitude are very close to one another and so you can walk through 15° in a few minutes. This means that places just a few miles apart can have a real time difference of many hours. In the polar regions however there is practically six months of daylight followed by an equally long night so it matters very little what time you make it out to be.

The other consequence was pointed out by the mathematician Charles Dodgson, better known as the author Lewis Carroll of the *Alice* stories, in the 19th century. As was just mentioned, if you travel east you are obliged to put your watch forward by one hour for each 15° of longitude that you cross in order to know the correct local time. Of course if you continue in the same direction, you will eventually return to your starting point, having advanced your watch one entire day. It follows that at some point on your journey there should be a line where, as you cross from west to east, the date suddenly reverts back to yesterday. We all may be used to the idea of the International Date Line now but when Dodgson used to recount this argument at dinner parties the distinguished guests would become very vexed indeed for what he was saying seemed quite preposterous. To their minds, the time of day was a fact of nature ordained by God, the Earth was a round ball going steadily around the sun and so the idea that there was somewhere a sudden jump in time was beyond belief. Besides, they *knew* that he was wrong! There was no such line. There was no place on Earth where it was Monday morning in one village while a few miles away they were still celebrating Sunday. All the same, his argument was irrefutable and it was not only Dodgson who was beginning to make this point. The famous French writer, Jules Verne, used the implicit date line jump to rescue his hero Phileas Fogg, the world's most punctual man, from losing his bet in *Around the World in 80 Days*. Fogg, after his many adventures, returns home to London one day late by his own reckoning, only to find the morning newspaper curiously bearing what seemed to be yesterday's date—Verne had used the anomaly to provide his tale with one of literature's cleverest twists.

It was true that in the 1800s there was no International Date Line but the need for one was beginning to make itself felt. Up until the 19th century the lack of a date line was unimportant. Huge tracts of the world were virtually uninhabited and the few people who were there had more pressing concerns than the calendar. Even if you lived in London it was of no concern what the date might be in New York until the era of instantaneous telegraph communications when suddenly it could matter what the time and date were in some far-distant land.

Of course any line of longitude could serve as our date line but since we placed the prime meridian at Greenwich, and measure time from there, it is natural to choose the same great circle, the other half of which is the meridian 180° east (or west) of Greenwich. This meridian passes mainly through the uninhabited regions of the Pacific Ocean so it turns out that placing Greenwich on the prime meridian was a good choice after all. It is true that for reasons of international convenience the International Date Line does not simply run along this meridian but has a number of kinks to accommodate the wishes of various island nations of the Pacific, and notably to keep Siberia and the Americas forever in different days of the week, but the modern arrangement is overall a sensible one.

Phileas Fogg's adventure raises the question as to what it actually means to go around the world. After all, we could imagine ourselves walking around the North Pole in a small circle in a minute. We would have crossed every meridian and returned to our starting point. Does this constitute a circumnavigation of the world? The answer is no, as we have stayed entirely in the northern hemisphere and have totally neglected the south. Any trip around the world must not only return to its starting point but must not be contained in any hemisphere on the globe. It should be a closed circuit that does not stay entirely one side of *any* great circle but rather crosses every one of them at some point. Another way of looking at this requirement is that it should *not* be possible to see the entire curve—no matter what the angle chosen to view the globe, some of the curve should be hidden from sight on the far side. With all due respect to Phileas Fogg, I think his claim must be doubtful on these grounds as his journey was northern hemisphere all the way, the most southerly point being Singapore, which lies just north of the equator.

The first true circumnavigation of the globe was that of the Portuguese navigator Ferdinand Magellan in 1519–22, or rather that of his surviving crew, as poor Magellan did not live to see the completion of the great voyage as he was killed in April 1521 when he

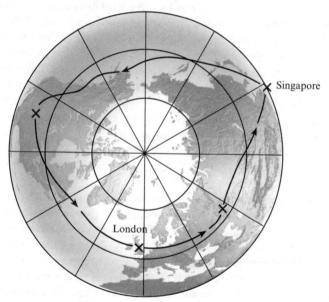

Fig. 1.4: Phileas Fogg's circuit

provoked the wrath of the natives of the Philippines through his insistence that they be converted to Christianity. Of the five little wooden boats that set sail from Seville only one, the *Vittoria*, returned three years later, after having sailed around Cape Horn, across the Pacific to the Spice Islands, and home again round the Cape of Good Hope and Africa. Magellan was the first to cross the Pacific Ocean although not the first European to gaze upon it as that privilege was accorded to Balboa some seven years earlier. Curiously, Magellan *was* the first man to sail around the world for he had visited the Philippines on a previous voyage and his return to Europe, followed by the leg of his 1519 expedition that took him around the tip of South America, over the Pacific, and *across* the Philippine meridian did constitute a complete rounding of the globe. Indeed that this should be so is a fitting tribute to a truly intrepid explorer. Five centuries on, it stands as a mighty feat that earns him enduring admiration down the ages.

No one would be inclined to argue that the voyage of the *Vittoria* was not a global circumnavigation yet it is not so clear what does and does not constitute a trip around the world. To qualify we might require that the journey pass through a pair of *antipodal points*. The *antipode* of a point P on the Earth's surface is the point Q exactly opposite P on the other side of the globe on the line through the Earth's centre. Every great circle through P also goes through its antipode Q and vice versa. For example the North and South Poles constitute one pair of antipodes. Since most of the world's land mass is concentrated in the northern hemisphere, starting in your favourite town, a tunnel bored right through the Earth to the antipodes would probably emerge in some deep ocean. There are none the less parts of China and Argentina that happen to be exactly opposite one another on the globe.

The requirement of passing through a pair of antipodal points will ensure that your circumnavigation does cross every great circle: no path that includes a pair of antipodes can be contained in one hemisphere as no antipodal pair of points, P and Q, can lie on the same side of any great circle. A great circle represents a shortest possible trip around a sphere. If the sphere is of radius r say, then its great circles all have a common length of $2\pi r$, where the number π is around 3.1416. If we have a closed curve C on our sphere the length of which is under $2\pi r$ then it can be shown to lie in one hemisphere. Certainly C is not long enough to contain two antipodal points. There is some subtlety here as this pronouncement does not remain true if the curve is not closed up, that is to say, it fails to return to its starting point. For example, take the curve C which comprises half of the equator of your sphere. This is only half of a great circle and so has length πr yet it does not sit within any hemisphere, that is to say, does not lie on one side of any great circle, because any great circle other than the equator itself meets the equator in two antipodal points, at least one of which lies in C. However, it would seem that a trip in which we traipsed around half of a great circle and then turned back and returned to our starting point by retracing our steps or following a line very close to the original should not count either as we seem to have deliberately avoided 'going around'. We might outlaw this type of path by insisting that every around-the-world voyage actually crosses every great semicircle (and not just every great circle) but this however represents an extremely tough condition. One can easily draw a closed path P that fluctuates a little around the equator for instance but if a slight bobble north of the equator is not matched by an equal bobble south of the equator at the antipodes then we can draw a great semicircle meeting the equator at right angles that you have missed. There are a number of natural questions about curves on the sphere that turn out not to have such clear-cut answers. Another is the question of just how long is a country's coastline.

Crinkly Coastlines and Convexity

Ireland is a tiny place compared to Australia and at first sight you might think that the coastline of the former is a tiny fraction of the latter. However the coast of Ireland is very rocky and bumpy compared to that of Australia, with myriads of little inlets, so if you were to measure it by winding a rope tightly around the shore it could turn out to be comparable to that of an entire continent. Indeed we can imagine a small island with thousands of deep but narrow glacial bays that could have a coastline of any given length while still maintaining a very small area as pictured in Figure 1.5.

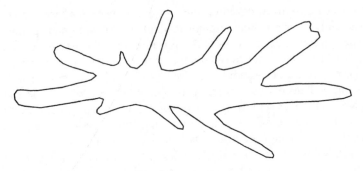

Fig. 1.5: A country with a long coastline and very little area

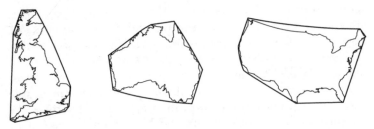

Fig. 1.6: Some national convex hulls

The correlation between the area of any island and the length of its coastline is a weak one. However, just picturing the globe again it is plain to see that there must be some other way of looking at this for the 'boundary' of Australia is enormous compared to the 'boundary' of Ireland. Our first measure of the boundary, the coastline, is too distorted by the fine detail to capture this idea so we must look to some broader measure of the size of the outside of the shape. This could be done by ignoring little inlets that we meet along the way as we measure the coast and cut across from one cape to the next. This viewpoint gives a construction known as the *convex hull* of the shape.

To make a model of the convex hull of the Australian mainland for instance you would begin with a cut-out map or a template of the island. Place the template inside a loop of cotton and draw it tight. The containing shape adopted by the cotton loop is the convex hull of the inside shape. The convex hulls of some well known countries appear in Figure 1.6. As you can see they are reminiscent of advertising logos in that they appear much as simplified or stylized versions of the real thing. The convex hull will be much more closely related to the actual size of the country in that a small country will still have a relatively short boundary for its convex hull. (At least this is true in practice, although it is easy to imagine a spider-shaped country, like that of Fig 1.5, that would have a small area while generating a large convex hull.) Convex hulls of figures make sense on a flat plane and also on the globe provided that the shape is not too big. However, when you draw an island on the sphere you divide the surface into two separate regions. In the plane there is always an *inside* and an *outside* to such a shape—the inside being bounded, that is to say of a limited size, while the outside is unbounded and stretches away indefinitely. This is not so on the sphere as what you deem to be the inside and outside is more arbitrary, especially when the curve is big enough so that the two regions into which it partitions the globe are roughly equal in size. If your shape sits in one hemisphere you will be able to draw your loop of cotton tight around it but not otherwise. For example, take Australia again but this time treat the Australian mainland as the *outside*!

For a shape to be *convex* means that if we join any two points within the shape to each other, the line (that is, a great circle path on a sphere) that we draw lies entirely within the figure. If that figure lies in a hemisphere then that hemisphere itself is convex and contains the original figure. The intersection of all the convex regions that contain the figure gives you the smallest convex region that does so, and so that it must be its convex hull. On the other hand if your figure is so large that it cannot fit into *any* hemisphere then the smallest convex region containing the shape is just the entire sphere because any convex region on the sphere with a boundary is contained inside some hemisphere: to see this imagine that you have a convex shape on the sphere and draw a great circle tangent to any point P on its boundary. The outline of the shape cannot leave the great circle at P only to return to it later at some other point Q as the entire great circle path from P to Q would have to lie in the figure precisely because it is convex. For this reason the figure lies entirely on one side of the great circle and so is contained in the corresponding hemisphere.

This idea allows us to suggest that a circumnavigation of the world is a closed curve C that is either a great circle or one whose convex hull is the entire world, for to say that the convex hull of C is the whole world amounts to saying that C does not fit into any single hemisphere of the globe, wherever you might draw your great circle.

Round Triangles

Triangles on the globe are not like the ordinary flat ones that you might be used to. If you draw a small one, say with sides no more that a few hundred yards long, it will appear to be much the same as those you met in your school textbooks. It cannot be quite the same though for what are the sides? The sides will not be straight lines but portions of great circles as these are, as we have been saying, the geodesics of the sphere—the shortest paths between two points. This has a rather curious effect on the sizes of the angles of the triangle. If these angles are measured very carefully you would find that they add up to something a little greater than 180°. The difference can only be small as your triangle is almost identical to an ordinary flat triangle whose angles always sum to exactly two right angles. However things are much more extreme for big triangles as you can see immediately by imagining a very large triangle. For example take the triangle formed by the equator, the prime meridian and the meridian 90° east. The meridians meet the equator at right angles (all of them do) and these two particular meridian lines meet each other at 90° also. This gives you a triangle with three right angles! Things can get even more extreme. You can make the meridians move apart until they are virtually separated by 180°—now the sum total of the angles approaches four right angles. Finally, by opening up the right angles at the equator until the triangle almost becomes the equator itself, we can increase the total angle measure up to a limit of $3 \times 180° = 540°$, that is to say, six right angles. This is as large a triangle as you can possibly draw on a sphere for a triangle is a convex shape and as such must be contained in some hemisphere.

To gain some idea as to why a spherical triangle has this angular excess and how it relates to the triangle's size it is best to remind yourself why the angles of an ordinary flat triangle sum to that of a straight angle, that is to 180°. Draw a triangle, choosing one side to call the base, and draw a line parallel to the base through the corner of the triangle not on the base as in Figure 1.7. The angle marked b equals the angle opposite to it, b', while a and a' are the same as are c and c' exactly because the line drawn is parallel to your base. It follows that the sum of the three angles of the triangle, $a + b + c$, is equal to $a' + b' + c'$, and the latter three evidently add up to 180°, the angle measure of the straight angle of the top corner of the triangle on your parallel line.

So much for flat triangles. Why does this reasoning fail when you draw your triangle on a sphere? You can certainly draw the same type of figure, taking the base of the triangle to be the equator say. The parallel line can also be drawn but it would be a parallel of latitude and not a great circle. For that reason the angle a will be *larger* than that of a'. For the same reason c will be larger than c' although b will equal b' the same way as before. Now a', b' and c' still add up to 180°, and so it follows that the sum of the angles of your triangle, $a + b + c$, will *exceed* 180°. Of course if the triangle is small compared to your sphere it will be very nearly flat and so the excess over 180° will be correspondingly small but for a large triangle, as we have seen, the excess is proportionately large.

How is this excess related to the area of the triangle? In as simple a way as possible: the excess is directly proportional to the area of the triangle. If you wish therefore, you can measure the area of the triangle directly by measuring its angular excess. For example, if one triangle has an excess of 10° and another on the same sphere has a 20° excess, then the second has twice the area of the first, even

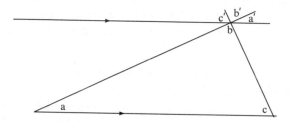

Fig. 1.7: Angle sum of a triangle

though it might be a quite different shape. The surface of a sphere of radius r is, for reasons explained in a Chapter 4, equal to $4\pi r^2$ so that, if we take the radius to be one unit, a hemisphere, which is the largest the area of a triangle can be, is 2π. The angular excess in this extreme case is $540° - 180° = 360°$. Since the area of a triangle is proportional to its total angular measure in excess of $180°$, the area of a triangle on our sphere with, for example, a $30°$ excess would be $2\pi \times (30/360) = \pi/6$ square units.

A novel consequence of this formula is the utter impossibility of having two similar triangles on a sphere; two different figures are *similar* if one is simply a larger version of the other and, in particular, two triangles are similar if, and only if, they share the same set of three angles. However, two similar triangles on a sphere would then have the same angular excess, and so the same area, forcing them to be *congruent*, that is to say identical in every way (except position). This is a very basic and elegant result but because such facts on the geometry of the surface of the sphere are not to be found in *The Elements*, the ancient texts of Euclid (*c*.300 BC), it seems to have been unknown until the time of Harriot, who produced a very pretty proof, although the equating of angular excess and area for spherical triangles often goes by the name of *Girard's Spherical Excess Formula* as that writer published his own proof in the book *Introduction nouvelle en algebre* in Amsterdam in 1629.[1]

A way of seeing the relationship between angular excess and area is as follows. Take a triangle T on a sphere and split it into smaller triangles, T_1 and T_2, by joining one corner to the side opposite (see Fig. 1.8).

The angle measures of T_1 and T_2 have respective values $180° + e_1$ and $180° + e_2$ say, where e stands for the excess of the total angle measure over that of a straight angle. Adding up the angle sum of T we see that it is the sum of the two angle measures of T_1 and T_2

---

1. The history of the subject is not cut and dried: Euclid did write an applied mathematics text on spherical geometry in relation to observational astronomy known as the *Phaenomena*; the fact that two spherical triangles with the same angles are equal is to be found in the work *Sphaerica* of Menelaus (fl. AD 100) which survived only through the translations of medieval Arabic scholars who were themselves key contributors to the development of both plane and spherical trigonometry.

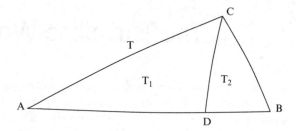

Fig. 1.8: Angle sum of a spherical triangle

That is to say

$$(180° + e_1) + (180° + e_2) - 180° = 180° + (e_1 + e_2),$$

minus that of the straight angle at the point D. And so we see that the excess of the angle measure of the big triangle is just the sum of the excesses of the two smaller triangles of which it comprises. In other words, just as the area of T is the sum of the areas of T_1 and T_2, the angular excess adds in the same manner and that is why the area of the triangle is proportional to the excess. This idea of measuring area by angular excess can then be extended to figures with four or more sides by partitioning spherical polygons into triangles along the lines used in Euclidean geometry.

2 ○ The Travelling World

Dusk is the time when we are reminded that our world is adrift in space. As the sun sets, the brightest stars appear as faint and distant points of light piercing the twilight sky. As darkness settles the stars brighten, mysteriously they seem to draw nearer and, especially if the night sky is clear and without the moon, we are left standing beneath a heavenly sphere glowing with thousands of stars, gloriously bright. Away from city lights it is the most sublime sight in nature and in witnessing it we can understand why peoples the world over have imagined the night sky as the home of the gods.

The ancient belief that all the heavens revolved around the Earth was certainly naive and is often regarded as arrogant, placing, as it does, Man at the Centre of the Universe. This impression is re-inforced by the past actions of the Roman Catholic Church which insisted that the geocentric universe of Aristotle be accepted as dogma. The Inquisition suppressed the emerging truth of the helio-centric (sun-centred) model of Copernicus (1540) and forced Galileo (1633) to recant his own beliefs along these lines, despite their being founded on hard evidence and sound argument. All the same we should bear in mind that it is the most natural thing in the world to presume that the heavens circle the Earth. This leads us to regard the earth as the centre of all but, at the same time, the belief that often accompanied this outlook was one where the earth was the *lowest* of all places where we humans were condemned to spend our mortal lives while our spirits yearned to join the divine and eternal world of the heavens above. This is not arrogant but altogether natural and uplifting. What is more, this seems to have been a universal outlook among all the peoples of the world. It is to be noted however that nomadic peoples, for instance Australian Aborigines, evolved cultures based on respect for all of nature, while the city dwellers of Europe believed their world to be a wretched place full of misery and baseness from which they longed to be delivered.

The stars of the night sky are often referred to as the *fixed stars*, a term that stands in need of explanation as, to a casual observer, they seem far from fixed and move in a variety of bewildering ways. Let us discuss this. First, they appear to revolve in the sky as the night progresses but that at least is to be expected due to the spinning of the earth on its north–south axis. This axis points to the celestial poles around which the stars appear to revolve during the night. These are the points where the line of the Earth's axis appears to pierce the celestial sphere. Those stars close to the polar directions such as Polaris, the Pole Star, in the northern hemisphere and the stars of the Southern Cross in Australia are affected relatively little by this rotation. They never set below the horizon and are visible every night from suitable locations. However, if we look out on a February evening we see the mighty constellation of Orion, the Hunter, who, with his starry belt, stands proudly in the heavens, but is nowhere to be seen in the middle months of the year. In what way are the stars of Orion fixed?

All stars are extremely distant (we are not including the sun, moon, and planets under this heading) and the motion of the Earth around the sun and the motions of the stars themselves are negligible in comparison to the enormous spaces between them. Because of this, the direction from the Earth to any star is practically fixed and this means, for example, that at any given spot in the world the stars of Orion's belt will always rise in the same place every day—but not at the same time each day. We cannot see Orion in July because during that month he lies beyond the sun, that is to say the Earth no longer lies between Orion and the sun as it does in February but its orbital travels have placed the sun between us and the constellation of the Hunter. Orion rises and sets during the daytime when the stars are invisible so that constellation cannot be seen from anywhere on the planet during this season.

In order to grasp the overall picture we can imagine what we would see if we could turn the light of the sun off for a day. If we stood at one of the poles during our day of darkness the stars above us all would revolve once parallel to the horizon. None would rise or set and the stars of the other hemisphere would be invisible to us below the horizon. This is what is experienced in any case during the long polar night. On the equator, however, the twenty-four hour night would be one unlike any ever experienced, for we would see all the stars of the firmament revolve about us, although any particular star would rise and set, only being visible for half of the day. The stars near the plane of the equator, which include the stars of the zodiac, would rise in the east and fly over our heads to set in the west, while constellations near the direction of the Earth's axis, such as the Great Bear, would creep reluctantly above our horizon (in the north in the case of the Ursa Major), always remaining low in the sky. If we stood at an intermediate latitude we would experience something in between. We would see most, but not all, the stars in the heavens; most of the stars we see would not be visible all the time but would rise and set at some time during the day although some, such as Polaris if we were north of the equator, would be visible the entire time, indeed the Pole Star would appear, as always, quite stationary with everything turning about it.

Next to the daily rotation of the Earth about its axis, the feature of its motion that most affects our lives is the tilt of that axis. If the axis of our planet were perpendicular to the plane of its orbit there would be no seasons at all. Night and day would be of equal duration everywhere all year long and weather patterns would undoubtedly be quieter, although the equator would still be much warmer than the poles. Indeed the whole idea of the year would all but vanish as the annual revolution of the earth about the sun would only be observable through the motions of the stars which, though intriguing, would not seem to tie in with anything experienced on Earth, and so would never have taken on the importance that they have. However, the tilt of over $23°$ is very substantial, causing massive seasonal differences, especially away from the equatorial regions, which are built into the very fabric of life on Earth.

Not all the variation in the cyclic motion of the heavenly bodies is however connected with the axis tilt as is revealed when we ask ourselves the question:

How many times does the Earth rotate in a single year?

There is a little trap in this question as the answer is not $365\frac{1}{4}$, which is the total number of solar days in a year: we see 365 sunrises each year while for one year in four there are 366, which we allow for through the invention of the Leap Year. The Earth however spins one more time than this, a fact that has nothing to do with the seasons but, rather, is geometrical.

Even if the axis of the Earth were perfectly upright, the *sidereal day*, the day of the stars, would be about four minutes shorter than the twenty-four hours between successive sunrises. To see why, look at Figure 2.1 which shows the position of the Earth from above the North Pole, N, as it orbits the sun on two successive days. (The angle swept out by the line joining earth to sun has been exaggerated to make the following point more visible.) The lower picture represents the position of the Earth after it has rotated on its axis exactly once, and so, from the viewpoint of an earthbound observer, the fixed stars have revolved precisely once overhead. In the first picture, the observer P is experiencing sunrise at this moment but, due to the orbital motion of the Earth around the sun, one sidereal day later, the same observer is still waiting to see the new dawn. Before the sun rises again for P the world must turn through a further angle equal to the angle swept out by the line joining Earth and sun the preceding day. This is of course a small angle: since there are just under $365\frac{1}{4}$ days in the year the angle in question is under one degree. The length of the day is 24 hours which equals $24 \times 60 = 1440$ minutes. This makes the time between the lengths

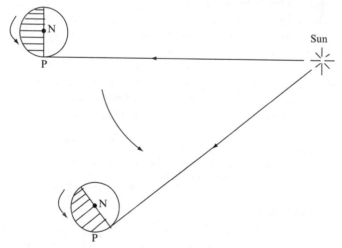

Fig. 2.1: Difference in solar day and the day of the stars

of the solar and sidereal days equal to $1440/365\frac{1}{4}$ minutes, which is, to the nearest second, 3 minutes and 57 seconds.[1] Although there are $365\frac{1}{4}$ solar days each year, the Earth spins on its axis $366\frac{1}{4}$ times during that period.

Passing to another point, knowing the number of days in the solar year is crucial if you are to get your calendar right. A 365-day calendar will advance one day every four years ahead of the true pattern of the seasons and in the course of a lifetime one would witness a drift of several weeks that would certainly be noticeable and disruptive to planting and harvesting. After seven centuries the seasons would seem completely topsy-turvy. The calendar established by Julius Caesar was based on a $365\frac{1}{4}$ day year. It provided good service for centuries despite some peculiarities that still persist and are now bizarre anachronisms. Instead of beginning our year at, say, the spring equinox or the solstice, the first day of our year is 1 January, in order to coincide with the convening of the Roman Senate.[2] February has two fewer days than every other month just to satisfy the vanity of Roman Emperors: July, named after Julius Caesar, had its length increased to 31 days, and then Augustus Caesar insisted on the same treatment for his month, August, to show he was no less a figure than Julius. To satisfy these conceits two days were stolen from February.

The absolute length of the solar year is a matter of debate. It can be taken to be the length of time for the sun to reappear in exactly the same position in the tropical sky but that depends (very slightly) on which point in the sky you begin measuring from. None the less, however you measure it, the year is a little shorter than $365\frac{1}{4}$ solar days and the error of the Julian Calendar was bound to catch up with us as time went by. By the 16th century the Julian Calendar had

---

1. This difference shows some variation throughout the year as the length of the solar day varies because the angular speed of the earth around the sun is not quite constant due to the earth orbiting the sun in an ellipse rather than a perfect circle; around Christmas the earth moves *faster* as it is a little *closer* to the sun and so the corresponding *increase* in angular speed manifests itself in a solar day that is some seconds *longer* than in June.

2. It seems we owe a debt of sorts to the Emperor Nero: Tacitus in his *Annals of Imperial Rome* informs us that, in AD 55, 'The Senate had decreed that future years should begin in December, the month of his birth. But he retained the old religious custom of starting the year on January 1st.'

slipped about ten days behind reality and so Pope Gregory the Great introduced the Gregorian Calendar that is used today and cut ten days out of October 1582 for good measure to place Easter back where it belonged. Every year divisible by four has an extra day, 29 February, *except*, and this is the Gregorian reform, those years divisible by 100 which are not leap years *although*, as a final piece of fine tuning, years divisible by 400 do have the leap year day. In accord with this, the years 1600 and 2000 were leap years although 1700, 1800, and 1900 were not and nor will be the year 2100. The Gregorian Calendar thus has three fewer days every 400 years compared to the Julian and this approximation is accurate enough not to require further adjustment for many millennia.

Let us now think about the way our planet happens to be tilted over. It could have been worse—the planet Uranus has an axis set totally awry so that it rides around its orbit almost on its back so to speak. If the Earth had an axis tilted at 90°, what would we experience?

The polar circles would extend right down to meet each other at the equator. Everywhere in the world would have a full day of sunshine at the summer solstice when the axis pointed directly at the sun, and a totally black day six months later at the winter solstice, alternating, of course, between the northern and southern hemispheres (see Fig. 2.2). At each equinox, when the axis of rotation was tangent to the orbit, we all would experience twelve hours of daylight and darkness but even near the equator this balance would not prevail for long and in much of the world the swing back to near total daylight when heading into summer and very short days when pointing away from the sun would be the norm.

The poles would see tremendous weather extremes, being subject to the continuous blazing light of the overhead sun in summer, while in winter we would have to endure total cold blackness with no hint of sunlight. The equatorial regions would be strange twilit worlds. The sun would rarely set and would linger most of the time around the horizon. As each equinox approached there would be a brief few weeks when the sun would suddenly swing high into the sky to pass directly overhead at midday on the equinox only soon to retreat low into the sky of the other hemisphere where it would revert to its prevailing sulky behaviour, clinging close to the far

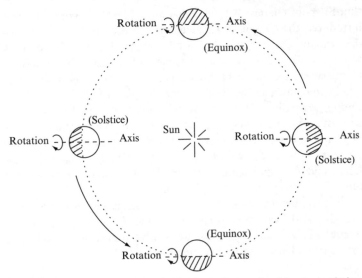

Fig. 2.2: Seasonal differences of planet with 90° axis tilt. At each solstice the hemisphere facing the sun has constant daylight with the sun directly overhead at the pole; the other hemisphere suffers the opposite extreme

horizon, equally reluctant to rise as to set. The weather at the equinox, already noted for its stormy character on the real Earth, would be incredibly violent in this severely tilted Earth as the atmosphere was forced to adjust to the wild changes in heating patterns occasioned by the passing of the equinox.

The changes generated by the $23\frac{1}{2}°$ tilt of the axis are remarkable and severe enough. At the June solstice the northern half of the globe is tilted towards the sun. Any point P $23\frac{1}{2}°$ south of the north pole will enjoy twenty-four hours of sunlight on this day, as can be seen in Figure 2.3, for even at midnight the sun is just visible although it never rises higher than 47° above the horizon at any time. In general, the maximum angle of the midsummer sun if you stand at latitude $a°$ outside of the tropics is $(90 - a)° +$ axis tilt and this leads to a qualitative difference between a summer's day in the tropics and that of, say, northern Europe. The equatorial regions offer little or no shade in the heat of the day but no matter how warm it may be on an August afternoon in Siberia, the trees of the forest cast their long

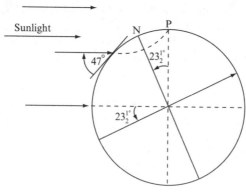

Fig. 2.3: Midsummer sun

shadows, as if to remind the visitor of a winter that is never too far away.

The latitude $90° - 23\frac{1}{2}° = 66\frac{1}{2}°$ north, marks what is called the *Arctic Circle*. Within this circle is the Land of the Midnight Sun. Of course, all northerly regions experience very long days at this time. Various factors tend to extend partial daylight in any case—the width of the sun's disc, refraction from below the horizon which allows the sun and its light to penetrate even after it has 'set', and diffusion of light through the atmosphere from the sun below the horizon leaving behind twilight.

Even in the latitude of southern England, twilight persists all night at the end of June so that the sky never completely darkens. At the poles, there is only one 'day' per year but even here there is a bias in favour of daylight as the sun is above the horizon for 189 days and below for 176. Twilight itself tends to be long at the poles where the sun is never very far from the horizon and short at the equator where the sun sets almost vertically so that all light vanishes remarkably quickly after sunset while sunrise is a swift and bursting spectacle.

Points above the equator at a latitude equal to that of the axis tilt experience the sun directly overhead at midday on the summer solstice (again see Fig. 2.3). Along this line, known as the *Tropic of Cancer*, the afternoon offers no shade at all from the hot sun directly overhead. Equally far south, there lies the *Tropic of Capricorn*, named after the sign opposite the Crab in the zodiac, where the sun passes directly above in late December.

On the equator itself, each day is of the same duration all year around, yet there are some seasonal variations none the less as the sun is directly overhead at the equinoxes but lies to the north in the northern summer and in the south in the northern winter. Any region in the tropics, that is to say one lying between the Tropics of Cancer and Capricorn, experiences the sun directly overhead on two separate occasions either side of the summer solstice. For instance, in Thailand, which lies in the northern tropics, the hottest sun comes after the northern spring equinox every April.

The Story of the Motions of the Sun, Moon, and Planets

The great astronomer and mathematician Johann Kepler finally established the *heliocentric* (sun-centred) model of the solar system as a fact when he published in 1609 three laws of planetary motion. Up until the beginning of the 17th century, the basic motions of the solar system were still matters for debate. Kepler worked under the great Danish observational astronomer Tycho Brahe who persisted in favouring an Earth-centred model of the solar system in which the planets circled the sun but the moon and sun still orbited the Earth. However, through stubborn perseverance based on Brahe's observations, Kepler eventually showed that what is observed is consistent with the planets, including Earth, orbiting the sun in ellipses rather than circles with the sun at one focus (plural 'foci': a term explained shortly).[3] What is more, his Second Law states that the line joining the planet to the sun sweeps out equal areas in equal time intervals so that the planet moves more quickly when close to

---

3. Aristarchus of Samos proposed in the 3rd century BC that the earth was a sphere in orbit around the sun. His brave conjecture was seriously considered but rejected partly because the theory predicted that this motion would lead to a change in direction of the fixed stars throughout the year, yet none could be observed. Aristarchus suggested, rightly as we now know, that the lack of visible 'parallax' as this effect is known is simply due to the stars being enormously distant.

the sun and slows down when at the far end of its orbit: in Figure 2.4 the three areas denoted *ASB* swept out by the orbiting planet in equal time intervals are themselves equal in area.[4] Finally, a peculiar connection was proved linking the length of the planet's year, *T*, with its average distance from the sun, *D*, in that Kepler's Third Law states that the quantity, T^2/D^3, is constant for any planet so that T can be determined from *D* and vice versa. This final rule is peculiar and surprising in its precision and the fact that it involves comparing a square and a cube. Kepler had no theory leading to this conclusion—the rule was found empirically to be the one, out of many that Kepler tested, that worked.

That Kepler was prepared to spend years on tedious, fruitless calculations with little prospect of success is sometimes cited as evidence that, despite his outstanding achievements, he had a mediocre brain with his only exceptional ability being dogged determination. However, some of Kepler's mathematical work was a precursor to the integral calculus, while his writings on projections and tessellations seem much in advance of his own age. It is true that some of his notions were fanciful and turned out to be wrong but he none the less proved to have an exceptional mind that achieved great things.

Kepler's Laws represented a break with the past as everyone up until this time seems to have accepted that the underlying motions

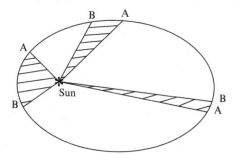

Fig. 2.4: Kepler's Second Law

---

4. Later generations of physicists would view this as an instance of the principle known as *conservation of angular momentum*: the phenomenon by which area is swept out at a constant rate is a feature of a body orbiting another subject to any kind of central force.

of heavenly bodies were necessarily circular as this appeared more in harmony with the idea of the Music of the Spheres. Ellipses had not been considered in their own right before.

A circle is the figure drawn when you fix a drawing pin on a sheet of paper to act as your centre, place a loop of string around it, put a pencil in the loop, and, while holding the loop taut, move the pencil around. The pencil then draws a *circle*, the set of all points a fixed distance, (the *radius*), from your centre. If we replace the single tack by a pair of tacks and carry out the same procedure we draw an oval shape known as an *ellipse*. The centre of the circle has now been replaced by two points, F_1 and F_2, known as the *foci* of the ellipse and, as you draw your curve, the loop of string forms a triangle as shown in Figure 2.5. The sum of the lengths of the three sides of this triangle is always the same, being the length of your loop of string. One side is fixed, being the distance between the two foci so that the *sum of the lengths of the other two sides is fixed also*. An ellipse then turns out to be the set of all points whose *sum* of the distances to the two foci is some fixed number. It is this number that replaces the diameter of the circle so that the ellipse is a generalization of the idea of a circle.

Ellipses are to be seen all around—the appearance of a tilted circle such as the rim of a cup is an ellipse and ellipses appear when you cut a cylinder or cone by a plane that is not parallel to the base (a parallel cut yields a circle in both cases). We shall have more to say about these remarkable curves later in the book.

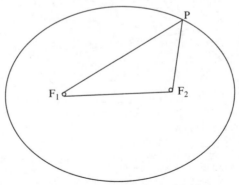

Fig. 2.5: Tracing an ellipse with a thread taut around two fixed foci

Kepler did not give reasons why his laws should be true, rather he merely demonstrated that they fitted the facts. Indeed, it would have been hard to imagine what kind of reasons you might seek to account for these rules, so the fundamental question of why the laws should prevail remained unapproachable for a further eighty years until the time of Isaac Newton. The mystery would no doubt have persisted much longer had not Newton been able to frame the Universal Law of Gravitation and, possessing the powerful mind that he did, develop a whole new system of mathematics, the calculus, from which it can be shown, among many other things, that Kepler's Laws are a consequence of the so-called Inverse Square Law of gravitational attraction.

The history of physics and the attendant mathematics of this era is not simple and it is not all Newton. During the 17th century other scientists were toying with ideas close to that of the Law of Gravity and, as a quite separate matter, calculus, the mathematics of rates of change, was at an incipient stage. The great French amateur mathematician, Pierre de Fermat, played with ideas related to calculus, as did Kepler himself to a limited extent. The great rival and contemporary of Newton, Gottfried Wilhelm Leibniz, developed the calculus himself and published results before Newton. It was Newton however who ended this nebulous state of affairs by establishing modern physics in his *Principia Mathematica* in 1687. This book represents one of the great turning points in our understanding of the world and nothing that went before can compare. Up until this time, a great classical or medieval philosopher might have conversed on a broadly equal footing with any scientist throughout history. However, neither Aristotle nor Thomas Aquinas would have known where to begin with Newton. Even Archimedes would have been stunned.

It would not be until the twentieth century when once again the prevailing fundamental physical problems of the day would be resolved, at a stroke, by a single individual when Einstein established his theories of relativity. We are left to speculate just how much longer we would have had to wait for the breakthrough without the two men concerned. In each case, all the necessary ingredients for progress were in the air so to speak, so we can argue that it was inevitable that, before too long, others would have stepped into the breach and filled the gap. On the other hand, a protracted and debilitating period of confusion may have ensued in which the best minds of the day despaired of ever making progress, declaring the difficulties insuperable, and advising others to turn their backs on what appeared to be a hopeless state of affairs. We are not to know.

Frustratingly, Newton still did not show the world all the techniques he had at his disposal even in the *Principia*, for he eschews his own new methods and insists on establishing his results in the published work through use of classical mathematics, a fundamental ingredient being the theory of conic sections (ellipses, parabolas and hyperbolas) developed by Apollonius of Perga, a younger contemporary of Archimedes, in the third century BC. Perhaps he was more concerned that his work be seen to rest upon absolutely firm foundations rather than to risk displaying all his new mathematics which, while making the explanation more natural and comprehensible, could have left him open to criticism of lack of rigour.

Everyone seems to have heard the story about Newton discovering the Law of Gravity after being struck on the head by a falling apple but fewer people seem to know what that law is. If asked, you might meet with the reply, 'what goes up must come down' or such like, as if to suggest that Newton was the first person to notice that things fall down. No, the Universal Law of Gravitation says that between *any* two objects there is a mutual and equal force of attraction. The great psychological hurdle to clear was the realization that the force applies universally to *all* objects and not just to special heavenly bodies which are no longer to be regarded as special at all. Even Kepler seems not to have explored this line of thought, although Galileo, who discovered the four principal satellites of Jupiter, mountains on our own moon, and new general principles of the mechanics of projectiles, was surely moving in this direction.[5] What is more, this force of gravitational attraction can be quantified and is governed by a simple rule. The force is proportional to the *product*

---

5. The greatest mathematician of the 14th century, Nicole Oresme of Paris, deserves mention here as he deduced by graphical techniques not widely used until the 17th century, that the distance travelled by a uniformly accelerating body is proportional to the square of the time. As with many other of his ideas, such as the use of fractional indices to represent roots of numbers, he was well in advance of his own era.

of the two masses and *inversely proportional* to the square of the distance between them. This law is entirely reasonable once its meaning is explained.

Suppose we have two masses, m and M, with m the smaller. They exert equal forces on each other but because a given force will cause a proportionately greater acceleration in a smaller mass than a larger one, the effect of M on m is more visible than that of m on M. If M were replaced by a body of twice the mass, $2M$, then the force on the little mass would double, which is surely what we should expect. On the other hand, the further the objects are apart, the weaker the force they exert on each other. The force being inversely proportional to the square of their separation means that if we double the distance between m and M, then the force of attraction drops to $(1/2)^2 = 1/4$ what it was previously and if the distance separating them increases by a factor of 3, then the force drops to $1/9$ what it was before, and so on. This kind of tailing off of influence is very common in the physical world and is consistent with the way we might imagine the influence of gravity to behave for it is saying that the behaviour of gravitational attraction is akin to the way light intensity decreases with separation. To explain, suppose that we have a candle which is steadily emitting light. Picture the light source to be sitting at the centre of a series of transparent spheres, all with the candle at their common centre. Since the surface area of a sphere is proportional to the *square* of its radius (explained in detail in Chapter 4), a sphere of twice the radius of another lying inside it will have four times the area of the inner sphere. However, the same amount of candlelight passes through both spheres, so that the intensity of the light at any point on the outer sphere must be only a quarter that for a point on the inner sphere. It follows that if gravitational attraction radiates from a source in the same manner as light from a candle, we would expect that its influence would diminish in this inverse square manner. This is Newton's Law of Gravitation and from it Kepler's Laws and the general motions of the stars and planets can be understood and predicted. In particular, the elliptical orbit of a planet around the sun could be explained.

It is not strictly correct to say that one object orbits another as in reality the object pair orbit around a common point. In the case of a planet orbiting the sun or indeed the moon orbiting the Earth, the common point around which they revolve actually lies within the larger body so that the smaller object seems to do all the orbiting while the larger object simply has a slight wobble. It is this wobble that modern-day astronomers use to detect otherwise invisible planets around stars.[6]

Newton's Law of Gravitation applies to *point masses*, that is, masses that are so small compared to the space separating them that they can be regarded simply as points. For the case of something like the Earth–moon system it would not be satisfactory to consider each body to be a mere point. However, both bodies are close to spherical and Newton did prove that, under certain mild assumptions, a body in the vicinity of a massive sphere experiences a gravitational attraction that is identical to what it would feel if the mass of the sphere were all concentrated at its centre. Since spheres are highly symmetrical bodies this is perhaps not a totally unexpected result but it is a very convenient conclusion to have at hand when studying celestial mechanics and one that is proved using calculus.

What would happen then should an object, such as a satellite, be placed several hundred miles above the Earth? The answer is, it would behave as did Newton's apple, and would fall down with a crash. If the satellite has any speed at all away from the line joining it to the Earth's centre, it will enter into orbit around the centre of the Earth in an elliptical path. There is still trouble however, in that if the speed of the satellite is small, the ellipse will be very flat and tight and the path of the orbit will go through the Earth itself—in other words it will still fall to ground. If the speed is sufficient to avoid attempted travel through the Earth (and this must include the Earth's atmosphere) then the orbit will be stable and the satellite will orbit indefinitely without the need to expend further energy. The higher is its orbit, the less speed the satellite requires to stay up. It so happens that at a distance of about 23,000 miles above the equator, the satellite completes one orbit in exactly one day so that it stays above the same place on the Earth's surface. A set of three such

6. As did 19th-century astronomers: Adams in England and Leverrier in France, each unknown to the other, detected and predicted the position of a new planet, Neptune, from the wobble in the orbit of Uranus. Using only pencil and paper they calculated the position of the new planet which was duly discovered in 1846 when the great telescope in Berlin was trained on the spot.

geostationary satellites as they are called forming the corners of a triangle in space is what is required to have a stable global satellite communications network that can reach all points on the Earth's surface at all times.

So then the orbit of the Earth around its parent star is an ellipse. It is however not a very elongated ellipse and a scale drawing of the Earth's orbit from above could easily be mistaken for a circle—the distance between the Earth and the sun varies between 91 and 94 million miles and so the intensity of warmth we receive from the sun does not change much on this account. From the viewpoint of the Earth though, it is the sun that seems to revolve around the Earth during the course of the year and this apparent annual orbit of the sun against the background of the fixed stars is known as the *ecliptic*. The rotation of the Earth is very stable and so forms a good basis for a clock. If we try to mirror this rotation with a sundial (an invention of sunny Arabic climes) we find that we have a clock that runs fast or slow depending on the time of year, although these rhythms are balanced and so a sundial periodically returns to the right time throughout the course of the year.

There are two causes of the drift in solar time seen through a sundial. One is due to the *eccentricity of the earth's orbit*: since the Earth moves faster in January (the angular velocity increases over 3 per cent above average) the solar day is about eight seconds longer at the beginning of the year than in the northern summer months. In the course of three months a sundial accumulates an error of about eight minutes due to the Earth's orbit being an ellipse and not a circle. The second and more substantial source of error is due to the axis tilt as felt through the tilt of the ecliptic to the plane of the equator. A movement by the sun of one degree along its path does not correspond to the same angle in the shadow of the sundial which causes a sundial to gain or lose up to 20 seconds a day due to the inclination of the ecliptic. Both these factors link together to bring about a predictable rhythm in the pattern of error which leads to an English sundial, for instance, being over 16 minutes fast in early November and over 14 minutes slow in mid-February; the exact relationship governing these measurements is known as *the equation of time*, and, as with anything involving shadows and angles, is expressed in terms of trigonometric functions. If you trace out the curve defined by the midday position of the shadow of the tip of your sundial throughout the seasons you will find it gives you a figure-of-eight variously referred to as a *lemniscate* or *analemma*.

The five visible planets, since they orbit in almost the same plane as the Earth, are also to be seen in the vicinity of the ecliptic. However their behaviour is otherwise very erratic, not at all like the fixed stars, and indeed the very word planet means wanderer. These wanderings remain entirely unfathomable if you adhere to a geocentric universe but the underlying explanation is simple enough once we place the sun at the centre of the solar system.

First there are the two interior planets, Mercury and Venus, whose orbits lie within those of the Earth. Since Earth never lies between Venus and the sun, it is not possible for Venus and the sun to make an angle of 180° with the Earth the way the other exterior planets of Mars, Jupiter and Saturn regularly do. Both Mercury and Venus must always be either morning stars or evening stars in that, from the viewpoint of an earthbound observer, they are never too far from the sun and so can only be visible around dawn and dusk. Indeed the maximum angular separation of Venus from the sun, which is just in excess of 45°, allowed Copernicus to calculate the distance of Venus to the sun in *astronomical units*, one unit being the distance of the Earth to the sun. This can be done by simple trigonometry or equally accurately by a scale drawing as seen in Figure 2.6. At maximum angular separation, the line from Earth to Venus is tangent to its orbit, making the angle *EVS* a right angle, so the unknown side *VS* turns out to be about 0.72 astronomical units: Venus is about $93 \times 0.72 = 67$ million miles from the sun.[7]

The sun always and the planets generally drift eastwards around the line of the ecliptic as the year progresses. However, the planets can exhibit a retrograde motion where for a time they move westwards in the sky against the backdrop of the fixed stars before reverting to the prevailing eastward direction. This is most striking with the planet Mars which is the nearest planet to Earth that lies

7. Copernicus calculated the distances of all the five visible planets to the sun and the lengths of their respective years to within 1% of their true values: an astounding advance in the first half of the 16th century before the advent of telescopes.

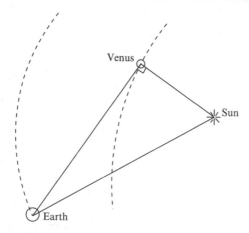

Fig. 2.6: Comparing the distances to the sun of Earth and Venus

outside of our orbit. This fickle behaviour was considered the greatest problem in astronomy around the time of Brahe and Kepler but the general principle becomes transparent enough once we accept the heliocentric model of Copernicus. The retrogression of this planet is as much the fault of Earth as Mars as it is entirely due to the faster moving Earth overtaking Mars on the inside as depicted in Figure 2.7.

The motion of the moon as seen from the Earth is very complicated, being significantly affected by the gravitational pulls of both the sun and the Earth, and is important as the moon is the chief bringer of the tides. The very word *month* is derived from the moon. The moon revolves once around the earth and returns to the same place in the sky in 27 days, always keeping the same face towards us so that the Earth is permanently invisible from the far side of the moon. Because of various wobbles in the Earth–moon system, about 60 per cent of the surface of the moon can be observed from the Earth at one time or another. The astronauts of Apollo 8 were the first human beings to gaze upon the far side of the moon at Christmas time in 1968 although photographs from space probes were previously available. However, because of the revolution of the Earth about the sun, the period of *lunation*, the time from one full moon to the next, is 30 days. (This is the same geometric phenomenon that

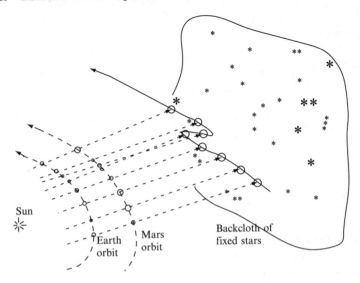

Fig. 2.7: Sighting lines indicating apparent retrograde motion of Mars

leads to the Earth's solar day being longer than its sidereal day, as explained earlier.)

When the far side of the moon is in sunlight, the side facing us is in darkness, and so the moon is *new* when it is between us and the sun. As the month progresses, a crescent appears with the horns of the moon always pointing away from the sun. At full moon, the Earth lies somewhere between the sun and moon and, if the Earth intervenes directly between them, the moon is eclipsed for a time and lies in the Earth's shadow. An eclipse of the moon occurred during the Peloponnesian War of the 5th century BC, and the Athenian writer and general Thucydides took time out from recounting battles to record the incident and offer the scientific observation that such an event seemed only possible at full moon, and not at any other phase, although he does not commit to print any thoughts he might have had as to why that should be.

The waxing and waning of the moon through its phases is of course due to the changing portion of the side of the moon facing us that lies in daylight. One prediction of the Copernican model of the solar system was that interior planets should also exhibit phases as, for example, the night-time side of Venus is facing us when it lies between Earth and the sun. This prediction was first verified by Galileo with his telescope for the phases of Venus are not visible with the naked eye. Although Galileo did not invent the telescope, the invention coming from Holland, he built a series of telescopes of increasingly better quality for himself for the purposes of astronomy and made exceedingly good use of them, discovering the four major satellites of Jupiter and the phases of Venus.[8] His discoveries of terrestial features on the moon such as mountains together with blemishes on the sun itself (sunspots) did much to undermine the classical view of the heavens as a realm of perfection whose nature was of a totally different cast to that of our lowly Earth.[9] It now seems from Galileo's notes that he even observed the outermost giant planet Neptune in the early hours of 28 December 1612, centuries before its official discovery in 1846. On two further occasions the following month he again observed this faint 'star' and went so far as to suggest that it had slightly altered position. Whether it ever crossed his mind that he was gazing upon a new planet we shall never know. The five visible planets had been known since antiquity and astrological myths had been spun around their supposed roles in our lives since time immemorial so it would have been a staggering step to announce in 1612, on such scanty evidence, that there was, after all, another one. The first discovery of a new planet was that of Uranus which was found by William Herschell while viewing from his back garden in Bath on the night of 13 March 1781. Uranus lies inside the orbit of Neptune and it is said to be just visible to some very sharp naked eyes. Both planets are similarly sized giants, being much larger than Earth although much smaller than Jupiter and Saturn. Deep space probes have recently revealed Uranus as a featureless ball of gas but Neptune, in contrast, turned out to be a most exquisite planet, coloured a beautiful shade of blue.

---

8. James Harriot did likewise in England at about the same time and made a similar series of discoveries including spotting moons around Jupiter.

9. This naive and ignorant view persisted however: just prior to the discovery of the first asteroid, Ceres, on the opening day of the 19th century, the renowned philosopher Hegel upbraided the scientific community for searching for such things, for he had 'proved' by philosophical insight alone that the existence of another planet was absolutely impossible.

Returning to the motion and appearance of the moon, although the precise motion of the moon is very complicated, it does lie close to the line of the ecliptic so that the full moon in particular occupies the same place in the zodiac as does the sun six months later. One facet of the full moon that never ceases to surprise is the almost alarming proportions it seems to take on as it rises. This illusion, for it is an illusion, seems to be psychological rather than optical for the width of its disc is no greater than when it is high in the sky. In our minds we seem to compensate and perceive such objects near the horizon as large. Despite the romantic associations of moonlight, the moon is often regarded with suspicion. The very word lunacy seems to connect it with madness and even Shakespeare could find the moon threatening, darkly observing that she 'comes nearer the earth than is her wont'.

Certainly the Bard was right to suggest that the very proximity of the moon means it directly influences the Earth, as is testified by the tides. Earthly tides are the combined effect of the gravitational pulls of the moon and sun on our oceans with the moon the main instigator. Although predictable, tidal motion is very complicated. At any one moment there are several tidal waves sweeping across the oceans of the globe and some places can experience up to four tides in one day. However, the basic pattern is of two tides per day that are caused by the drag of the moon on the Earth as it rotates below. This might suggest an underlying daily cycle of tides rather than the twice-a-day phenomenon that we commonly witness but the ocean is trawled twice as it passes below the moon, once as it faces the moon and a second time as it passes under the moon on the far side twelve hours later. For example, one of the principal tidal waves is generated across the Pacific and begins on its eastern rim when that ocean spins under the moon as the earth revolves from west to east. From the viewpoint of the ocean, it experiences much the same drag as the American West Coast again leads the ocean under the influence of the now invisible moon half a day later, again retarding the rotation of the water from west to east, and leaving behind a second tidal bulge for the day.

A much rarer but more spectacular effect of the moon is its periodic eclipsing of the sun. This is somewhat a freak of nature as the phenomenon owes its impact to the coincidence of the apparent sizes of the lunar and solar discs. Although there is some apparent variation caused by the relative distances between these three bodies not being constant, the disc of the moon just covers that of the sun at totality. If the moon were a little smaller or more distant, no totality would ever occur. The transit of the moon across the face of the sun could still be observed but there would be little apparent dimunition of daylight as totality is required for this dramatic effect.

Indeed Venus attempts to eclipse the sun on occasion but its disc appears so small it is an event that would normally pass unnoticed. It is in any case very rare—transits of Venus occur in pairs separated by eight years, but the passage of time between pairs is 113 years. There were no Venusian transits in the 20th century, the next being on 8 June 2004. The first correctly anticipated and observed transit of Venus was by Jeremiah Horrocks of Lancashire on 4 December 1639 just prior to sunset on an otherwise cloudy Sunday afternoon and the stated purpose of Captain James Cook's voyage in which he discovered the East Coast of Australia was to observe the 1769 transit of Venus from Tahiti.[10] (Note that 'rounding up' December 1639 to 1640 we see that $1769 = 1640 + 113 + (2 \times 8)$, in accord with the previous comment about the timing of transits: Horrocks's observation was evidently the first of an eight-year pair while Cook's Tahitian sighting was the second of the 18th-century pairing.)

In contrast, you cannot fail to notice a total eclipse of the sun if you happen to be standing under it. If the size of the moon's disc were larger, the duration of totality would be longer but we could not see what astronomers most cherish which is the chance to directly observe the surrounding corona of the sun and other phenomena associated with the sun's boundary that would be obscured by a larger or closer moon. As totality approaches the sun appears to go through phases as the disc of the moon visibly slices into that of the sun. However, since our eyes adjust to fading light, none of this prepares the observer for the sudden onset of totality when, in a moment, the daylight sky turns to darkness, the stars come out, and the sun is replaced by a black disc with coronal halo. This

---

10. The West Australian Coast was discovered by Dirk Hartog in 1616 and there are suggestions that other Dutch merchantmen may have landed even earlier: Australia was originally christened New Holland.

awe-inspiring experience was desperately frightening for people of earlier ages, whether or not the eclipse was anticipated. Such a shock could bring men to their senses as is famously retold in connection with the eclipse of 28 May 585 BC (predicted by Thales of Ionia some say) which happened during the battle between the Medes and the Lydians in Asia Minor. Both sides took the eclipse as a sign of utmost heavenly displeasure and suddenly became eager to conclude peace.

As to be expected, we also hear tales of exploitation of eclipses by high priests and the like who knew the secret of their prediction in order to gain power over the peoples of the Earth through apparent power over Heaven. Not all religious leaders however have behaved so unscrupulously. It is said that when the people, terrified by an eclipse, approached the Prophet Muhammad in confusion at this terrible omen that coincided with the death of the Prophet's own son, he explained that it meant nothing, that it was a harmless natural event, and that they should go home and not worry about it.

3 ⃝ The Geometric Picture

The two sides of the mathematical coin, discrete mathematics based on counting, as opposed to the continuous mathematics that arises through measurement, still form the basis of modern mathematics and the tension between the two branches persists today. The nature of this tension first emerged clearly during the lifetime of Pythagoras in the 6th century BC for it was at this time, in ancient Greece, that arithmetic and geometry reached the level of sophistication required to formulate the difficulties.

The Pythagorean era is often called mathematics' Heroic Age where big questions were posed by mathematical pioneers who had little with which to tackle them except their raw wits. The confident philosophy of *All is Number* that began the era had soon to be abandoned and the Greek mathematics that emerged was more based on the attitude that *All was Geometry*.

Measurement and Geometry

An early use of geometry and measurement arose in Egypt where the ancient Greek chronicler Herodotus, often granted the title the Father of History, tells us that the Nile's annual flood regularly washed away boundaries and landmarks so that a system of accurate measurement was needed in order to reaffirm who owned what: indeed the Greek word *geometry* means 'Earth measure'. A general fascination with pure geometrical shapes is a feature of any civilization and the Great Pyramid of Cheops (2600 BC) is a fine testament to this. Yet the founder of geometry was Thales of Ionia (640–546 BC) who is said to have impressed the Egyptians by calculating the height of the Great Pyramid through use of shadows and similar triangles. Like his famous successor, Pythagoras of Samos (*fl.*520 BC), he was of Phoenician origin and is reputed to have travelled widely in the East. Classical Greek geometry was the fruit of seeds that passed from Asia to Europe in the 6th century BC.

The story goes that Thales found the height of the Great Pyramid by measuring its shadow at a moment when the length of his own shadow exactly coincided with his height and the sun's rays were at right angles to the pyramid's base (see Fig. 3.1). He inferred that the shadow of the the pyramid must then also be equal to its height and went from there. Although he could not measure the distance from the tip of the pyramid's shadow to the centre of its base, as that was concealed within the pyramid, he could measure the distance to the side of the base and add on half the length of the side to have an accurate estimate of the distance he wanted to know. This trick apparently had not occurred to the Egyptians of Thales' day despite the fact that their ancestors had built the pyramid and founded the art of surveying some 2000 years before.

No writings of Thales have come down to us yet there are a host of stories attached to his name and he seems to have been quite a character. He was mocked by a 'witty and attractive Thracian servant girl' known as Theodorus for falling down a well while trying to observe the heavens. However he could be worldly enough when he

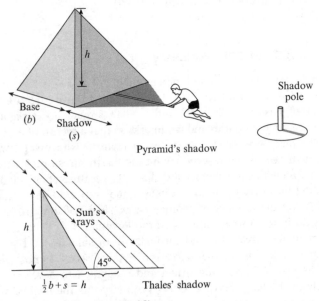

Fig. 3.1: Height of the Great Pyramid of Cheops

put his mind to it and he revenged himself on some who declared his philosophical musings worthless by shrewdly buying up the leases on all the local olive presses when he anticipated a bumper harvest and then hiring them out at a premium when they were in demand.

Leaving aside these tales we turn to the question as to why he earned the title of *Father of Geometry*. The reason is that Thales was the first to state and prove some of the basic facts of geometry. He made statements about *all* triangles, such as that the angles opposite the equal sides of an isosceles triangle are equal. (A triangle is *isosceles* if two of its *sides* are of equal length.) The geometric theorems of Thales are all on about this simple level and so it is fair to ask why we should be especially impressed by them. What Thales seems to have done is to start mathematics along the deductive road that it has followed ever since. Although he may not have progressed very far along that road himself, he was among the first to travel it and in so doing set mathematics in the direction of its destiny. We have grown accustomed, from a very early age, to being aware of general observations such as the fact that every circle is bisected by its diameter (also ascribed to Thales) and so it is hard for us to appreciate that Thales lived in a time when such a statement represented a new and sophisticated mode of expression. (Bear in mind that there were no paper and pencils available: figures were often scratched in sand.) A person of our own time who might think of themselves as knowing little or nothing about mathematics might never the less comprehend the kind of things Thales was getting at better than some of Thales' learned contemporaries. An example of a parallel development would be the way in which a modern person, even one who regarded themselves as having very little feeling for modern art, could be quite comfortable with a painting by Turner or Picasso, a painting that would have left artists of past ages feeling very ill at ease, for to them such images would seem strange, almost incomprehensible, lying quite outside their range of experience.

And then there is the matter of proof. Thales showed that his theorems were true in that they could be deduced from other facts that he regarded as simpler. Although Thales may not have developed his geometry in quite the thorough and systematic way that we find in Euclid some three centuries later, the deductive step had been taken. Thales did not simply make a series of unconnected general observations about geometrical figures but rather he showed that beginning with some truths we can infer others. This line of thought, argument from premises to conclusions, lies at the heart of mathematics and indeed all the sciences.

As a sample of the kind of things known around the time of Thales we can begin with the properties that determine a triangle. A triangle is known if we know the positions of the three corners, the *vertices*. It is however possible to find the vertices from other information. For example if we know the positions of two of the vertices then we have one of the sides and if we know the two angles of the triangle at these vertices then the triangle is known as it can be completed in only one way: we simply extend lines in the required directions from the two known vertices and the third vertex is found where those lines meet. Similarly if we know the lengths of two sides of a triangle and the angle that lies between them there is only one way that we can complete the triangle and so the triangle is fully specified by these data. The first specification is often written as ASA (angle-side-angle) while the second is the SAS (side-angle-side) specification (see Fig. 3.2).

A third way of specifying a triangle is to give the lengths of the three sides, a, b, and c. How can we construct the triangle from this information? We draw a side of the given length a and then, using compasses, we draw a circle of radius b centred at one end of our line, C, and another of radius c centred at the other end, B, as in Figure 3.3. These circles meet at two points, one above and the other below our line of length a, and either of these points can be taken to be the third point of our triangle.

We see that our construction has been generous enough to provide us with *two* triangles with sides of the specified lengths so it is fair to ask whether or not they are the same. This in turn raises a

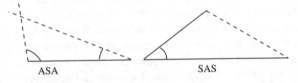

Fig. 3.2: Triangles determined by ASA and by SAS specifications

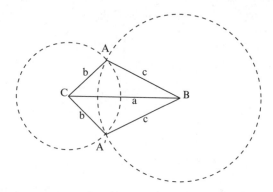

Fig. 3.3: Triangles with given sides

philosophical point. Two things can never be absolutely the same for otherwise it would be impossible to tell them apart. When we say that two things are the same what we really mean is that they are identical in all ways that are of current interest. This of course depends on context. For example, when playing cards two aces may be regarded as the same (meaning of the same value) in one game where suit does not matter but in a different game one could be stronger than the other and so the two cards would then be treated as distinct.

There is a word for being the same in geometry and that word is *congruent*. In particular we say that two triangles such as $\triangle ABC$ and $\triangle A'BC$ in Figure 3.3 are congruent. However, there is more to say as even here the triangles are not as identical as they might be. To make the point consider the pair of triangles in Figure 3.4. The two triangles shown are congruent—the first can be slid on top of the other. More precisely we can transfer the first on to the second by rotating the first 90° anticlockwise and then translating the triangle, that is to say moving every point of the shape an equal distance in the same direction.

The same cannot be said however of the two triangles in Figure 3.3 for to pass from one to the other we need to flip one triangle over: we have to *reflect* one triangle in the line of their common side to pass from one to the other. This involves an operation of a type not needed in Figure 3.4 and one which is somewhat more

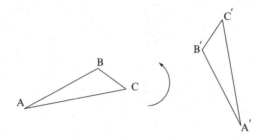

Fig. 3.4: Triangles congruent without reflection

complicated as, if you imagine the triangles as flat templates, in Figure 3.3 one triangle has to be *lifted* out of the plane in which it lies in order to be placed on top of the other, whereas in Figure 3.4 the operations of rotation and translation can both be carried out in the plane of the triangles without the requirement of a third dimension in which to operate. In contrast no amount of sliding around of the triangles in Figure 3.3 can place one on top of the other because they have different *orientations* in the sense that $ABCA$ takes you around the top triangle in a clockwise direction while the corresponding list of points $A'BCA'$ in the lower triangle takes you around in an anticlockwise cycle. In order to reverse these orientations some kind of reflection is required.

Generally, if less information is given about the triangle then it is not uniquely specified. For example, given one side and one of the angles at one of its vertices the triangle can be completed in any number of ways as the side at the given angle can be taken to have any length at all and the triangle can still be completed. Similarly if one angle is given and the length of the opposite side is known there is no limit to the number of triangles that can be drawn. (This is a little less obvious: you might care to draw a sketch of your own.) If the three angles are given then that specifies the *shape* of the triangle but it can be drawn with any area you wish and still have the three angles demanded. Two such triangles are called *similar*.[1] (See Figure 3.5.)

---

1. It is important that we are working in a plane here for a triangle on a sphere is determined once its angles are given, as was explained Chapter 1.

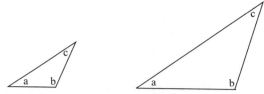

Fig. 3.5: A pair of similar but non-congruent triangles

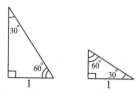

Fig. 3.6: Similiar yet different triangles with common side length

If the values of two of the angles are known you have been, in effect, given the value of all three (as the three sum to 180°) and if we know the length of one side then the triangle is determined provided we know the *position* of those angles relative to the side, but not otherwise as shown in Figure 3.6: we have two similar right triangles with acute angles of 30° and 60° that each have a side of unit length on the right angle yet the triangles have different areas.

There is another exceptional set of circumstances that gives rise to exactly two possibilities and for that reason this is known as the *ambiguous case*. It arises when we are given two sides of a triangle and one of the angles *not* included between the given sides. Figure 3.7 illustrates the two distinct possibilities that emerge in this case. The side *b*, angle at *A*, and the length *a* are given yet there are two distinct completions of the triangle, one acute (that is to say all of its angles are *acute*, meaning they measure less than a right angle) and one obtuse (meaning that the triangle has an *obtuse* angle greater than 90°; a triangle can of course have at most one such angle).

We are given the side *b* and the value of the marked angle at the vertex *A* together with the length *a* of the side opposite the angle at *A*. The third vertex can then be either of the points where the circle centred at the other end of *b* whose radius is of length *a* meets the line drawn in the direction specified. The two triangles that offer a

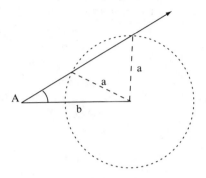

Fig. 3.7: Side-side-angle does not determine a triangle uniquely

solution to the specification problem are very different but both may be constructed using a straight edge and compasses as we have just demonstrated. This pair of tools, a straight edge (not a ruler with an attached scale but just an edge) and compasses for drawing circles were regarded as fundamental in classical mathematics and the demonstration of the existence of a geometrical object was taken to mean a way of constructing it using these tools alone. To paraphrase Euclid, a line may be drawn from any one point to any other (using your edge) and any line may be extended indefinitely while a circle may be drawn (by your compasses) with any point as centre and with any given radius. For this reason straight edge and compasses are often referred to as the *Euclidean tools*. We shall have more to say about them in Chapter 7 both as to their power and their limitations.

Finally we return to the point alluded to earlier: if two triangles are similar and a pair of their *corresponding* sides are equal then the triangles are equal (or rather congruent). A glance back at Figure 3.6 will confirm the significance of the word *corresponding* in this statement. There we have two similar triangles that are not congruent. It is true that each triangle has a side whose length is one unit but these are not corresponding sides: in the *larger* triangle the unit side is opposite the *smaller* (30°) angle whereas in the smaller triangle the unit side lies opposite the larger acute angle of 60°. It is the *vertical* side of the triangle on the *right* that corresponds to the side of length 1 in the triangle on the *left* for that is the side opposite the smaller of the two acute angles. The fact that the non-corresponding sides happen to be of equal lengths is of no importance.

If one triangle is compared to a similar one, say by placing the smaller insider the larger as in Figure 3.8, then the ratio of corresponding sides is the same for all three pairs. For instance, if AB' is twice the size of AB then the same is true of AC' and AC, and $B'C'$ is similarly twice BC. This is not a difficult aspect of geometry to accept but it is one with far-reaching applications, both in theory and practice. One device that relies on it is the use of the so-called *diagonal scale* that allows measurements to be made by eye to an accuracy of 1/100th part or indeed 1/200th part of an inch.

Diagonal scales were often marked on rulers up until the 1960s but now seem to have quite gone out of fashion. None the less they work as well as they ever did and have not been replaced by anything better. A picture of one can be seen in Figure 3.9.

In order to construct lengths specified to 1/100th part of an inch we use the diagonal scale (the distance between 0 and 1 on the scale represents 1 inch) and a pair of dividers, which is a tool consisting of two identical arms fastened together at a moveable joint like a pair of compasses, except that each arm ends in a sharp point. When the dividers are opened up to some specified length, that length can then effectively be lifted and accurately marked elsewhere on the page. Now, for example, if we want to mark a length of 1.47 inches we first count up seven places on the vertical scale, corresponding to the second decimal place of our measurement, and place one point of the dividers at the intersection of the seventh horizontal with the vertical line through the 1 inch mark, as shown by the small circle on the right in Figure 3.9. Now open the dividers and place the point of

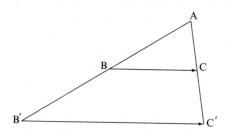

Fig. 3.8: Superposition of similar triangles

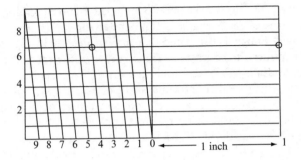

Fig. 3.9: The diagonal scale

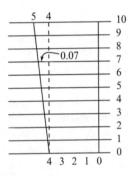

Fig. 3.10: Similarity and the diagonal scale

the other arm on the same horizontal at the point where it meets the diagonal numbered 4 (corresponding to the first decimal place) as shown in Figure 3.9 by the small circle on the left. The points of the dividers are now separated by exactly 1.47 inches.

Why this works is all down to similar triangles and is shown in Figure 3.10. Joining the point 4 by a vertical line to the corresponding point on the top parallel we have a triangle the vertices of which we have marked 4, 4, and 5. The vertical side 4–4 is divided into ten equal segments by the parallel lines of the scale, giving us a series of similar triangles, each inside the next, which have segments of the lines 4–4 and 4–5 as their sides and a segment of one of the parallels as base. Since the points marked 4 and 5 are separated by 0.1 inches the lengths of the parallels between the lines 4–4 and 4–5 are $\frac{1}{10}, \frac{2}{10}, \frac{3}{10}, \cdots, \frac{10}{10}$ of 0.1 inches. That is to say, these lengths are respectively .01, .02, .03, . . . , .10 of 1 inch. The diagonal scale uses the sides of similar triangles to guide the user to the correct placements. Its effectiveness is due to the gently sloping diagonals that allow a small horizontal difference to be magnified into a larger vertical difference that is discernible to the naked eye.

The Pythagoreans

Pythagoras is the pivotal character in the history of mathematics and he undoubtedly led a most interesting and colourful life, although accounts of it differ widely. He was born about 569 BC on the Greek Island of Samos where his father, Mnesarchus, was a merchant originally from Tyre.[2] Pythagoras seems to have travelled with his father to Tyre and to Italy and at about the age of 18 he met Thales, then an old man, in Miletus and was introduced to mathematical ideas through Thales' student Anaximander, although it was Pherekydes who Pythagoras regarded as his principal teacher. In 535 BC, like Thales a generation earlier, Pythagoras travelled to Egypt in search of wisdom but also perhaps to escape persecution at home in Samos where the tyrant Polycrates had seized power. The antiquity of Egypt greatly impressed the peoples of the Mediterranean and that civilization was regarded as the epitome of learning and wisdom. We find numerous tales of the sagacity and justice of the Egyptians in the writings of Herodotus. Undoubtedly the development of both Thales and Pythagoras owed something to the knowledge of Egypt, although this may have been of a quite general nature. It is unclear how much new mathematics Pythagoras learnt on his travels. His adventures were said to include becoming an Egyptian priest, and being taken to Babylon as a prisoner of war by Cambyses II of Persia, only to return to Samos in 522 BC where he founded his school of philosophy known as the Semicircle.

Pythagoras is sometimes compared with Buddha who lived around the same time and who also founded a religion, for that is

---

2. Bertrand Russell, in *History of Western Philosophy*, writes with characteristic wit: 'some say Pythagoras was the son of a substantial citizen called Mnesarchus, others that he was the son of the god Apollo, I leave the reader to judge between these alternatives'.

what the Pythagorean School became, with an underlying theme of the unity of nature and the divine spirit, with a particular belief in reincarnation. Pythagoras preached vegetarianism, that the Earth was a sphere, and that music could aid healing. Unlike Buddha, however, he also insisted that at the deepest level reality is mathematical in nature. This was a truly profound insight not generally shared with other mystics. It was this tenet that was the least well received by his fellow Samians. Indeed Iamblichus (writing some eight centuries later) says of Pythagoras' symbolic techniques for teaching mathematics that he had learnt in Egypt: 'the Samians were not very keen on this method and treated him in a rude and improper manner'.

And so Pythagoras left Samos for Italy in about 518 BC and founded a more successful school at Croton open to both men and women—Pythagoras himself married one of his students. Those who belonged to the inner order, the *mathematikoi*, led a regulated communal lifestyle, although there was an Outer Circle of the Society whose members lived apart and did not conform with all the rules on diet and personal behaviour.

Pythagoras left us no writings of his own and there is little in the way of personal description available to us. It seems certain though that he was a very talented musician, a player of the lyre, and put great store on loyalty to friends. Pythagoras travelled to Delos in 513 BC to nurse his benefactor and teacher Pherekydes until his death. Accounts of the death of Pythagoras himself are not consistent—he may have died as late as 475 BC and some descriptions have him living longer and indulging in more adventures. His school at Croton ran into difficulties and internal conflicts but survived, expanded rapidly after 500 BC, and became political in nature. It seems to have come to a violent end in Croton about forty years later. The Brotherhood, although scattered, persisted for perhaps another two centuries although the philosophical teachings of Pythagoras eventually passed into history. In contrast, his discoveries in music and mathematics live on. His theorem is still taught the world over, and always will be. Although not the way Pythagoras himself would have viewed it, Pythagoras' Theorem gives us an algebraic hold on the notion of distance and for that reason it lies at the heart of the mathematical study of the physical world.

However, the history of Pythagoras' Theorem does not begin with Pythagoras. The fact that the square on the hypotenuse (longest side) of a right-angled triangle was equal in area to the sum of the squares on the two shorter sides was appreciated by other peoples of antiquity well before the days of Homer and Classical Greece. For instance the ancient Babylonians, one thousand years before the birth of Pythagoras, were aware of the Pythagorean relation in the right-angled triangle to the extent of compiling astonishingly extensive tables of number triples such as (3, 4, 5), (5, 12, 13) where the sum of the squares of the two smaller numbers equals the square of the larger. A triple of this kind can be used to form a right-angled triangle all sides of which are whole numbers. The clay tablets contain triples such as (4961, 6480, 8161), which show that the Sumerians attached great importance to the Pythagorean relation and strongly suggest that they had a systematic way of generating these triples. There is indeed a formula for finding the entire collection of Pythagorean triples (there are infinitely many). The Sumerians could not have recorded such a formula in the modern fashion, where letters are used to stand for arbitrary numbers, yet might have known the recipe and how to apply it. (The Greeks themselves eventually found and proved the formula—it appears in Book 12 of Euclid's *Elements*, but here we are talking of the best part of two centuries after the time of Pythagoras.)

Although we cannot be sure, it seems likely that Pythagoras did not discover his theorem but came upon it in the course of his experiences. This does not make him a fraud for the great achievement of Pythagoras, or at least the Pythagoreans, was to prove the theorem in full generality. The theorems of the Pythagoreans also included the fact that the sum of the angles of any triangle is two right angles and more generally that a convex polygon with n sides has as the sum of its interior angles $2n - 4$ right angles while the exterior angles always sum to 4 right angles, corresponding to one complete circle. Again, these facts may have belonged to a corpus of geometrical folklore by the time of Pythagoras but were shown by his school to be consequences of much simpler facts and in this way they established geometry on a firm footing with scope for further development. They were inventing mathematics as we know it today and so to that extent the ancient school of Pythagoras is a part of the modern world.

The theorem itself is not only the most important, but the most proved of mathematical facts, with hundreds of distinct arguments recognized as leading to the celebrated conclusion and the authors are various and illustrious. For example there is a proof by the US President Garfield (1831–81) and another by Leonardo da Vinci (1452–1519) is of a particular style known as a congruency-by-subtraction. (For those interested, details of this one are given in the final chapter, 'For Connoisseurs.') It is not surprising that people keep discovering proofs, though often they turn out to be rediscoveries. For example, in 1873 H. Perigal devised a clever proof of the same genre as the proof by Leonardo that turned out to be known to the 9th-century Persian mathematician Thabit ibn Qurra.

Here is one demonstration based on similar triangles and general properties of areas. It makes use of the fact that if you double all the lengths of the sides of a triangle, then you increase its area by a factor of 4 or, more generally, if you multiply all side lengths of a triangle by n then the area is increased by a factor of n^2, that is to say that the area is proportional to the *square* of the length of the sides. Having said this, let us consider Figure 3.11.

The triangle $\triangle ABC$ is right-angled and we drop a perpendicular from the right angle A to meet the hypotenuse at D. Since the angles in any triangle sum to two right angles (the proof is on p. 15) we can deduce that the two smaller triangles into which the larger has been partitioned are each similar to the original. In detail the argument runs as follows: the triangle $\triangle ABD$ is right-angled and has an acute angle in common with the big triangle $\triangle ABC$ and so all three angles of $\triangle ABD$ equal the corresponding angles of $\triangle ABC$ and so the two triangles are similar. By just the same argument triangle $\triangle ACD$ is also similar to $\triangle ABC$ and so all three triangles are mutually similar. Now we invoke the general principle just stated that the area of each of the triangles is proportional to the square of the length of its hypotenuse and, since the areas of the smaller triangles add to give the largest, the same is true of the squares of their longest sides. This is Pythagoras' Theorem.

Another appeal of this proof is that it involves only the triangle itself. Many proofs are jigsaw-puzzle-type arguments based on comparing two different configurations of shapes built from several copies of the triangle and the squares on its sides. However the proof

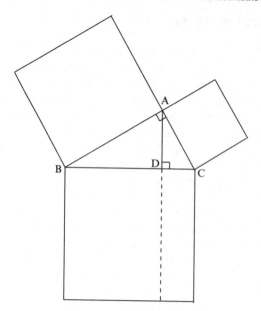

Fig. 3.11: A proof of Pythagoras' Theorem through triangle similarity

that is the culmination of Book 1 of *The Elements*, which is credited to Euclid himself, is a direct geometrical proof often known as the *Bride's Chair* because of the nature of the associated diagram. Although not the shortest possible proof it does establish some additional properties relating the triangle to the squares on its sides: in particular the result follows from the fact that the area of each of the smaller squares equals the area of the associated rectangle formed within the larger square when the perpendicular *AD* is extended as indicated by the dotted line in our diagram (see Chapter 8 for details). Whatever proof you prefer you should always ask yourself where the argument makes use of the condition that the triangle has a right angle because without that hypothesis the conclusion is not valid and so any legitimate proof must make crucial use of this special property at least once. In our proof we needed the original triangle to be right-angled in order that the smaller right-angled triangles into which it is split be similar to their parent triangle.

Crisis and Confusion

The maxim of the Pythagorean philosophy was that 'All is Number', by which it was meant that anything capable of precise understanding could be comprehended in terms of whole numbers and ratios between whole numbers. That four right angles made for one complete turning was the basis of geometry and the Pythagoreans studied numbers themselves and delighted in the many relationships they found between them. The confidence of Pythagoras on this score must have been bolstered immensely by his surprising discoveries in music. The story goes that he chanced upon these relationships by listening to the resonances that resulted from the weights of hammers used by a blacksmith. Pythagoras discovered that the interval of an octave is rooted in the ratio 2:1, that of a fifth in 3:2, that of the fourth 4:3, and that of a whole tone is 9:8. The Pythagoreans applied these ratios to lengths of a string on an instrument called a canon, or monochord, and were able to determine mathematically the intonation of an entire musical system. Music was taken to be more than just a pastime and was seen as a kind of divine voice. The very heavens themselves were thought to resound to the Music of the Spheres and so these discoveries were regarded as exceedingly profound.

Ironically it was the right-angled triangle itself that brought with it the seeds of destruction of the Pythagorean outlook for, as the founder himself is said to have discovered, a right-angled triangle the shorter sides of which are both of unit length has as hypotenuse the square root of 2, and Pythagoras proved that this number is *irrational*, that is to say there is no ordinary fraction, a/b, which, when squared, will yield *exactly* 2—that is a^2/b^2 cannot equal 2.

The simplest proof of this is a contradiction argument to be found in the works of Plato's student Aristotle.[3] To express it in modern terms, suppose to the contrary that a/b were equal to the square root of 2 and cancel any common factors these two numbers may have; in particular we may take it that at least one of a and b is odd, for if they were both even we could simplify the fraction further by dividing both top and bottom by 2. Then since $a^2/b^2 = 2$ it

---

3. For some interesting alternatives, see the final chapter.

follows that $a^2 = 2b^2$ and so that a^2, and so a itself, must be an even number allowing us to write $a = 2c$ say for some whole number c. Substituting accordingly we then arrive at $(2c)^2 = 2b^2$ which yields that $4c^2 = 2b^2$ and so $2c^2 = b^2$. But in the same way as before this now gives that b^2, and so by the same token b, is also an even number, contradicting that the numbers a and b have no common factor. This contradiction only arose through our assumption that the square root of 2 *could* be written as a fraction of two whole numbers and so it is this that must be wrong. We conclude, as did Pythagoras, that the square root of 2 is *irrational*.

This was serious. It contradicted the great man's philosophy and ruined some of the Pythagorean proofs of basic geometrical facts, for they had assumed that any two lengths a and b were *commensurable*, meaning that both were exact multiples of some suitable smaller length. To put this another way, it was taken as self-evident that, *in principle*, any constructed line could be measured *exactly* using a standard ruler, provided that we marked the ruler with a sufficiently fine scale. Pythagoras had proved this to be false for if $\sqrt{2}$ were commensurable with 1 then, by dividing the unit interval into some number, b say, of equal intervals we could measure the hypotenuse of the above unit isosceles right-angled triangle and find it to be exactly a say of these units of length $\frac{1}{b}$ which amounts to saying that $\sqrt{2} = \frac{a}{b}$. It is just this that the above argument demonstrates is impossible. You cannot measure the diagonal of a square with the same units with which you measure the side. However fine your scale, the tip of the diagonal will always lie between two of your scale marks. And not in the middle either—frustratingly we will always find it lying to one side or the other. We may find the idea of this annoying—to the Pythagoreans it was catastrophic.

This and other flaws in classical thinking were eventually resolved in the 4th century BC by Eudoxes with his *Theory of Proportions*, an account of which is to be found in *The Elements*, the classical text written by Euclid of Alexandria. Eudoxes introduced a theory that applied equally well to all length comparisons whether they be commensurable or not, by making subtle use of inequalities to deal with equalities. His approach was a model of mathematical purity, free of any taint of contradiction, and has therefore been admired down the ages by mathematicians as far in advance of what could have reasonably been expected at the time. Certainly no other contemporaneous civilization achieved anything comparable and the level of mathematical sophistication shown by Eudoxes was hardly bettered until the latter part of the 19th century. It may have been this level of achievement that the famous British mathematician G. H. Hardy had in mind when he described Greek mathematics as 'the finished article'. Its architects were to his mind kindred spirits—'members of another college' as he liked to put it—whose achievements were permanent, even more so than the authors of Greek tragedies, for languages die but mathematical truth lasts forever. Certainly the thirst for mathematical and scientific truth was real in the Academy of Plato in Athens in the middle of the 4th century BC and the extent that Eudoxes gave his heart to the subject is reflected in the quote attributed to him that he would 'willingly burn to death like Phaeton, were this the price for reaching the sun and learning its shape, its size, and its substance'. Other major figures at the time were Theodorus of Cyrene, Theatetus, the brothers Menaechmus and Dinostratus, and Autolycos of Pitane. Plato himself was a leading personality although historians of mathematics tend to be dismissive of Plato as a mathematician, regarding him as having more mathematical conceit than ability. However we cannot really judge—Plato certainly did much to foster the development of mathematics during this critical period and he may have made some direct personal contribution.

Despite its staggering sophistication, however, the Theory of Proportions had an artificial flavour that was difficult to apply in practice. (It is always easier to deal with things that are equal than with things that are not.) Greek mathematics never fully recovered its early confidence and continued to steer clear of mixing measurement and geometry. A modern student might label a side of a triangle by the letter a, thinking of a as a *number*, the length of a side, and then might engage in an algebraic calculation involving a. To Euclid such a symbol was merely a name for a *line*. Inhibition is never a good thing in the sciences and the Greeks' logical scruples contributed to the eventual stifling of classical mathematics. None the less it had a long way to run and the best was yet to come.

Euclid of Alexandria

Euclid is the most famous mathematical author of antiquity but his life is very obscure, although what little account of him as a man that has come down to us is most complimentary. It seems likely that he studied in the Academy of Athens under the students of Plato and came to Alexandria in Egypt around 300 BC. About thirty years earlier, the great University of Alexandria had been established by Ptolemy and it was to become and remain the major seat of learning in the classical world for centuries. It seems that about the time of Euclid it was run in a similar fashion to a leading modern university and that Euclid could be described as the world's first Head of a Department of Mathematics. He was an outstanding teacher and wrote a number of mathematically based works some of which, unfortunately, have been entirely lost. The most influential of these has been *The Elements*, a collection of thirteen books that formed the basis of mathematical instruction then and for more than the next two millennia. As an influential work therefore its longevity is unmatched. It would be wonderful to possess an original copy of *The Elements* but the modern versions are based on a revision by Theon of Alexandria prepared around 700 years after Euclid. In the 19th century, however, an older version of *The Elements* was unearthed in the Vatican Library that showed only minor differences from the edition provided through Theon.[4]

The content of *The Elements* is not only geometric but treats algebraic topics including solutions of quadratic equations, summing of geometric series, and elementary number theory including the *Euclidean Algorithm* for finding the highest common factor of two numbers through repeated subtraction. Never the less Euclid did remain wedded to the geometric style and so even common algebraic facts about numbers were demonstrated in what appears to us a strange and unnatural fashion through areas of geometrical figures. The books, such as Book 2, that contain much of this style of

---

4. Theon is also known as the father of Hypatia, a learned and beautiful young woman who wrote commentaries on the mathematician Diophantus. She was tortured and murdered in a most horrific manner by a fanatical Christian mob in the year AD 415.

argument have always been neglected in school mathematics for this very reason.

The Elements is none the less a work that has stood as the model of presentation up to the present day for it established the supremacy of deductive reasoning in mathematics.[5] All of Euclid's results were to follow from an initial sequence of postulates or axioms which were taken as self-evident. In doing this Euclid recognized that any system of reasoning must begin with some basic truths that are not consequences of earlier facts. The basic axioms none the less were taken as really true and his geometry was meant to describe the real world with the idea of points and lines corresponding to the physical ideas of dots on a page and thin straight strings. To this extent Euclid is not pure mathematics as emerged around the middle of the 19th century, when mathematicians took axioms to be genuinely unproved statements and studied the consequences when they happen to be true. A system of axioms could then be interesting in two distinct ways: because of the interesting consequences or because the axiom system could be regarded as the underlying framework of some real-world system.

The pure mathematical viewpoint of the 19th century[6] was a very liberating outlook as for centuries a great deal of energy was expended fretting over whether Euclid's Axioms were really true. In particular much heartache centred on the status of one particular axiom known as the Fifth Postulate. Could it be deduced from the others or at least be replaced by something simpler? Some of Euclid's postulates are completely general such as 'The whole is greater than the part', while others do apply specifically to geometry such as 'all right angles are equal'. Some are logically circular in that an undefined idea such as 'line' is defined as 'length without breadth' where two terms are introduced that are themselves not defined. This was not considered a real flaw for it was appreciated that Euclid was taking the nature of space for granted. No one was inclined to

---

5. And to an extent in all of science and philosophy—the reknowned 17th-century philosopher Spinoza wrote his own work in the style of a euclidean proof with axioms, deductive reasoning, and conclusion ending with QED as he and other philosophers yearned for a form of reasoning that approached the clarity and persuasive force of Euclid.

6. Betrand Russell attributes the 'invention of pure mathematics' to the Anglo-Irish mathematician George Boole.

argue about these statements as they matched everyone's experience of what could be said about two- and three-dimensional space.[7] The Fifth Postulate by contrast looks much more like a theorem than a basic truth. One version of the postulate is that in a plane,

Through a given point can be drawn a unique line parallel to a given line.[8]

This is the key assertion that gives Euclidean geometry its flavour. Some basic theorems do not require this postulate and Euclid is often praised for postponing the use of this contentious axiom for as long as possible—some results that have shorter proofs through use of the Fifth Postulate are proved by Euclid without invoking it. This is sound mathematical practice as it shows such results are particularly basic, being true independently of the Fifth Postulate. Nonetheless most of the important results such as Pythagoras' Theorem require the postulate for their proof. (Any proof of Pythagoras' Theorem needs the angle sum of a triangle to be that of two right angles: if you look at our proof on p. 15 of this result you will see that we made free use of the ability to draw a parallel line through a given point.)

I feel it best to pause in order to clarify how important all this once seemed to be, as it is not so easily appreciated by the modern mind since, for better or worse, there has been a change in attitude towards the importance or possibility of absolute truth in the past century or so.

Bertrand Russell epitomized the typical thinker of the late 19th century for his ideal was to produce a system of axioms that anyone would accept as incontrovertible from which all of mathematics could be derived. This aspiration was itself based on a false intuition as regards the nature of logic that he himself contributed to dispelling in the first half of the 20th century. Nevertheless the Euclidean ideal had a powerful hold on his mind. As a boy he was impatient

---

7. Indeed Immanuel Kant, perhaps the leading philosopher of the 18th century, took the Euclidean nature of space as self-evident and undeniable.

8. This form is known as *Playfair's Axiom*, named after the Scottish 18th-century mathematician John Playfair, but the axiom had been enunciated this way as early as the 5th century by Proclus.

with much of the humbug around him that he instinctively distrusted. He had heard that Euclid proved things and so was keen to be instructed in geometry by his older brother. His first lesson began with the axioms and so he was disappointed to learn that these fundamentals were to stay unproved. 'Why should I accept them?' he asked. His brother gave the only reply possible, that being, 'If you don't accept them, we can't go on.' Despite this initial setback he soon found mathematics irresistible, saying of Euclid that he 'had not imagined there could be anything so delicious in the world.'

One might think that such naive enthusiasm is the reserve of extraordinary individuals who were born to be mathematicians but a deep respect for Euclid and for mathematics in general was found much more widely among educated people. For instance, President Abraham Lincoln was a lawyer by trade who found himself leading a country embroiled in civil war, yet he turned to Euclid as a refreshing mental discipline. He boasted of his ability to demonstrate large parts of *The Elements* by heart and took a copy of the book with him on the circuit. Lincoln enjoyed the serenity, permanence, and clarity that the Euclidean world offered, together with the satisfaction of thoroughly understanding something of real substance. Another interesting insight from the same period came from Charles Darwin, the father of the theory of evolution, for his respect for the subject was based on the view that 'Mathematics seems to endow one with something like a new sense'.

Attempts to prove the Parallel Postulate as it came to be called inevitably ended up merely replacing it by something at least as questionable: three alternatives are (*a*) supposing that there exists a pair of lines everywhere equally distant from each other, (*b*) there exists a triangle whose angular sum is two right angles, and (*c*) a circle can be drawn through any three points not all on the same line. The most elaborate attempt to prove the parallel postulate was by the Italian Jesuit priest and Professor of Mathematics at the University of Pavia, Girolamo Saccheri, who, in 1733, published the book *Euclides ab omni naevo vindicatus*, Euclid Freed of Every Flaw.

Saccheri imagined a quadrilateral *ABCD*, as shown in Figure 3.12, where the angles at *A* and *B* were right angles and the sides *AD* and *BC* were equal. He then showed, *without using any version of the parallel postulate* that the angles at *C* and *D* were equal. The task

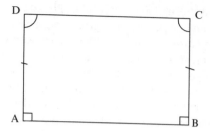

Fig. 3.12: The Saccheri Quadrilateral

before him was to show that they were not only equal but that they were both right angles without recourse to any version of the Fifth Postulate. He proceeded by contradiction. He first managed to gain a contradiction from the assumption that both angles C and D were obtuse (that is, greater than 90°).[9] He next turned his attention to the hypothesis that both angles are acute. From this he deduced consequences that today are recognized theorems of non-Euclidean geometry. Unfortunately he was not prepared to fully accept his own reasoning and insisted that he had found an inconsistency where in fact there was none. (He therefore claimed to have proved that the Parallel Postulate was a theorem of Euclidean geometry.)

Saccheri's somewhat misguided but ground-breaking struggle went largely unnoticed and it was not until a century later that the independence of the Parallel Postulate from the rest of Euclidean geometry was established. The great German mathematician, Carl Frederick Gauss, may have been the first to realize this but fell shy of publishing his own work so the credit goes to two men, the Hungarian, Janos Bolyai, and the Russian, Nicolai Lobachevsky, who independently took the same approach. They considered the Playfair form of the Parallel Postulate and considered the three separate

9. Well almost: to prove this he made a tacit assumption made by everyone up till that time, including Euclid, that a line may be *infinitely* extended as opposed to an endless extension—for example an arc of a great circle between two points on a sphere may be extended to a complete great circle which is endless without being infinite. Researchers continued to make life difficult for themselves as long as they persisted in not keeping their axioms logically separated from the intuitive model that their axioms intended to represent.

possibilities that through a given point can be drawn *more than one, exactly one,* or *no* line parallel to a given line. The third case can be eliminated *if you assume that lines can be extended infinitely,* which is the line of argument they took. They then went on to develop a non-Euclidean geometry based on the first case. Their work was not an immediate sensation as for one thing there seemed no need of these strange ideas. Neither man had provided a model for their geometries, that is to say, neither could point to a recognized setting in which their axioms held true. As long as this was the case it remained possible that Saccheri had been right and that this new set of axioms would lead to contradiction in which case all these new theorems would be meaningless. However, models of the new geometry were discovered first by Beltrami in Italy, and later Cayley in Britain, Poincare in France, and Felix Klein in Germany.[10] The brilliant German mathematician Bernhard Riemann showed that, by discarding the implicit assumption of the infinitude of 'lines' and some minor adjustments to the postulates, another consistent geometry can be developed from the hypothesis of the obtuse angle. The three geometries (Bolyai–Lobachevesky, Euclidean, and Riemannian) now go by the names of *hyperbolic, parabolic* and *elliptical* geometries respectively. Whether or not the true nature of space is Euclidean is a matter for physics, to be settled by observation and experiment, and not something to be decided by mathematics alone for we now know that it is not implicitly contradictory to suppose that the universe is non-Euclidean.

It is an irony of history that the axiomatic approach is not particularly suited to Euclidean geometry. It is in algebra and logic where this approach has been most successful—for example *group theory* is a major branch of abstract algebra that rests on just four axioms that form a very convenient starting point for the theory and are adequate to express it. In contrast Euclidean geometry is in reality based on geometric intuition acting within a partial axiomatic framework. Some of the claims in the Euclidean proofs cannot be

---

10. The Klein model is not intrinsically very interesting but has the virtue of simplicity—we take as our 'plane' a circle and 'lines' are taken to be finite line segments within the circle, with 'points' ordinary euclidean points: this geometry then obeys all the axioms of Bolyai and Lobachevsky showing that there is nothing inherently contradictory in their new geometry.

justified by appeal to the given axioms—some proofs implicitly assume that certain lines or arcs intersect which we now know cannot be deduced from the axioms alone but only by appeal to a picture or through some hidden assumptions. In principle an axiomatic geometry must be comprehensible not only to a blind man but to an entity with no experience of space, such as a computer. This can be done but it is a tall order. The number of axioms and primitive notions (terms not defined by other terms but known only through the rules that apply to them) is much greater than allowed for in traditional geometry. The strictly formal systems of Euclidean geometry that have been constructed typically have around twenty axioms and no particular collection of starting axioms is pre-eminent. A number of systems have been devised, each as good as the other. Although there have been attempts to introduce such systems into school geometry and so do Euclid 'properly' as it were, I would expect these systems always to be the province of specialists. They can have their uses as they do furnish ways of carefully and systematically ranking the status of theorems within a geometry and comparing one geometry to another. What is more, strict adherence to a formal system can alert researchers to blind spots in their analysis.

That it took two thousand years for mathematics to advance to the stage where serious criticism of Euclid was required in order to progress is testament to the enduring quality of classical mathematics. To be sure Euclid was implicitly criticized on presentational grounds even in the classical period—for example Pappus in the 4th century AD gives a proof of the Theorem of Thales that two angles of an isosceles triangle are equal which is slicker than that in *The Elements* and it has been observed that Euclid is often only as good as his sources and so the quality of the exposition is better in some of his books than others. There is though a patience in the work that comes from realizing that the subject-matter is so fundamental that it demands the greatest of care and thoroughness. For example, when Euclid postulates that a circle may be drawn (with compasses) with any given centre and radius, he means that when the compasses have been picked up the arms may collapse together again and so the drawer has no immediate trace of the given radius. In practice, modern compasses do not behave like this and can be used as a pair of dividers to mark a given distance elsewhere on the page. However,

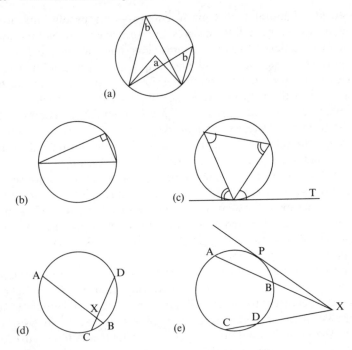

Fig. 3.13: Sample circle theorems. (a) The angle *a* at the centre is twice the angle *b* at the circumference. (b) Taking *a* to be a straight angle, in (a) gives that the angle in a semicircle is a right angle. (c) If T is a tangent to the circle we have equality of the angles as indicated. (d) Intersecting chords theorem: (AX)(XB)=(CX)(XD). (e) Intersection secants theorem (AX)(BX)=(CX)(DX)=(PX)^2.

this would be to postulate another tool and Euclid goes on to show how this is not *necessary* in that the action of marking a given distance with dividers can always be simulated using just a straight edge and (floppy) compasses so that the additional tool of dividers is merely a practical convenience and not a theoretical necessity. (Indeed it is an interesting exercise, requiring some ingenuity, to show that we can, for example, mark the length of a given line segment onto another using only Euclidean compasses and edge.) Modern compasses lend additional speed and simplicity and without them even quite basic constructions soon take on the appearance of a bewildering cobweb of lines. All the same they do not afford us any more power—any geometrical construction that can be done with the additional tool of dividers can, in principle, be done without them.

This kind of analysis is necessary in order to fully understand both the power and the limitations of formal tools. The ancient Greeks were rewarded for the respect they showed their own subject through the enduring influence of their work.

The other facet of Euclid which adds to its appeal is the simple beauty of the results. The theorems on circles and tangents have a charm that never fades and even today, when so little is proved or explained in school mathematics, they still appear in our textbooks. We close with a sample of a few (Fig. 3.13). For those who wish to be reminded of the proofs or perhaps to see them explained for the first time, the demonstrations are recorded in the final chapter.

Although surprising, all of the circle theorems are based on the fact that the angles in a triangle sum to two right angles and that a triangle in a circle two sides of which are radii is isosceles. The final pair of results is particularly pretty but each comes down to observing that certain triangles are similar. The theorems are then used by Euclid to justify the claims made for many of his geometrical constructions, some of which will be highlighted in Chapter 7.

4 The World of Archimedes

The greatest mathematician of antiquity was Archimedes of Syracuse. He lived to the age of 75, having spent the final years of his life frustrating the Roman efforts to take his city during the Punic Wars through the use of a series of ingenious mechanical weapons that some maintain included 'burning mirrors' that set fire to the wooden Roman ships. When Syracuse finally fell in the year 212 BC the story goes that Archimedes was murdered by an angry Roman soldier. Archimedes did have his final wish respected though, that his tomb should have inscribed upon it, for the benefit of future generations, a diagram of a sphere inside of a cylinder of the same radius and height. This was reported by the Roman orator Cicero who discovered the tomb in Sicily around 150 years later. Finding it fallen into disrepair, Cicero had the tomb of the world's greatest scientist restored. No trace of it remains today however and even the site of the grave is lost.

Archimedes spent some time in Alexandria, was probably taught by students of Euclid, and stayed in contact with leading figures there, yet he was born and died in the Greek colony of Syracuse in modern-day Sicily. Archimedes was the first mathematical physicist and, apart from weapons of war, invented practical mechanical tools such as the Archimedean screw, a system of helical pipes fastened to an inclined axle designed to raise water from the Nile to irrigate the surrounding countryside. None the less he regarded all this as secondary to his fundamental discoveries of general principles. His books *On the Equilibrium of Planes* and *On Floating Bodies* were the first real works of mechanics.[1] The Principle of the Fulcrum, that the turning effect or *moment* of a body placed on a lever is equal to the product of its weight times its distance from the pivot had been known much earlier but Archimedes' explanation and exploitation of the principle was more well founded and effective both in theory

---

1. Aristotle, a century earlier, wrote eight works entitled *Physics* but these were non-mathematical speculative efforts which, although interesting and influential, were very flawed.

and practice. The First Principle of Hydrostatics is due to him personally:

Any solid lighter than a fluid will float in such a way that the weight of the solid will equal the weight of the fluid displaced; a solid heavier than the fluid will sink but its weight in the fluid will be lessened by an amount equal to the weight of the fluid displaced.

The discovery of this principle of buoyancy is the basis of the 'Eureka' story for solving the problem of proving that King Hiero's crown was not pure gold.[2] However, even in such works as *On Floating Bodies* Archimedes was keen to turn to theoretical problems involving, for example, the behaviour of paraboloids of revolution.

In his treatise *On the Sphere and Cylinder* Archimedes discovered a beautiful relationship between the sphere and the cylinder that allowed him to unlock various secrets of the sphere including its surface area and volume and we shall explain these a little later. That he should place such store upon this relationship indicates that he believed it to be an aspect of mathematics that would arise time and again over the ages. For an example we can return to our airline map of the world first mentioned in Chapter 1.

A flat world map where lines of longitude as well as the lines of latitude are displayed as if they were parallel results from a *cylindrical projection*. The simplest cylindrical projection arises by taking your globe of the world (a sphere), placing it in a cylinder as on Archimedes' tomb, and projecting each point on the globe horizontally from the axis on to the containing cylinder (so the poles alone miss out) (see Fig. 4.1). To get your map you now need only slice the cylinder vertically along the projection of a convenient meridian, usually that 180° east, and lay your sheet out flat to have a nice picture of the world suitable for global viewing (Fig. 4.2).

The resulting image gives us a pretty reasonable idea of where places are in the world in relation to one another. Any portrayal of the surface of a sphere as a flat object must have distortions and

---

2. The Roman architect, Vitruvius, asserts that Archimedes showed the crown to be a fraudulent mixture of gold and silver simply by comparing the densities of gold, silver, and the crown by measuring the displacements when equal weights of each were immersed in a vessel full of water.

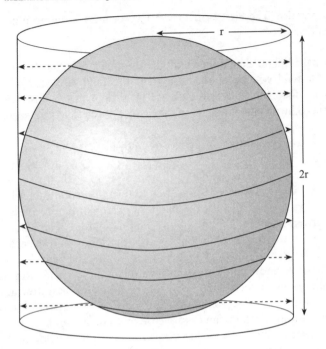

Fig. 4.1: Cylindrical projection

this is no exception. For example, the continent of Antarctica is spread out along the entire bottom edge of the map, and indeed the scale is only accurate at the equator.

Is the map useful to navigate by? If we are in London and draw a straight line on our map to San Francisco does that line relate to something usable on the real globe? As explained in Chapter 1, that straight line is not a great circle and so does not represent the shortest path between the endpoints of your journey. However, travellers of old, sailors in particular, were less concerned about the shortest route to their target destination than they were about hitting the target in the first place, so that finding the rhumb line, the bearing that would take them to their final port of call, was what was required. However, the straight line that you have drawn on your simple cylindrical projection is not the rhumb line either. This

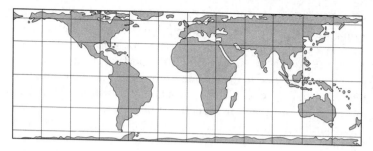

Fig. 4.2: Cylindrical equal-area projection

projection is not much use for serious navigation. The trouble is that it distorts angles. What was required was a flat representation of the world which preserved angles (the mathematical word for this is *conformal*) for then a straight line on the map between two points would represent the rhumb line between those points on the globe. The solution to this problem was provided by the Flemish cartographer and mathematician Gerardus Mercator in 1569.

Mercator, who introduced the word *atlas* to the world, was born in 1512 in what is now Belgium and was taught from 1530 by the mathematician Gemma Frisius. Frisius himself seems to have been the first person to point out that it was possible, in principle, to measure longitude using an accurate clock. Determining one's latitude at sea was relatively easy by measuring the height of the sun at midday with an astrolabe. Frisius pointed out that if, at the same time, you had a clock which was still on the time of some standard line of longitude then your longitude could be quickly calculated on the simple basis that one hour of time equals 15° of longitude. He was aware that 16th-century technology was not up to the task, adding that 'it must be a very finely made clock', and one that could stand the rigors of a sea voyage. Nonetheless, this was to be how the longitude problem that had so vexed seamen for centuries was solved. John Harrison's watch H4, in the sea trial to Barbados in 1764, proved up to the task and Harrison's timepieces had already been recommended by several sea captains, including James Cook, prior to this demonstration. Harrison's clocks all lay far in the future when Mercator and Frisius produced a globe of the world in 1541, at a time when some learned men were still insisting that the world was flat.[3] The conformal Mercator projection of 1569 is strictly not a projection at all as the image cannot be constructed simply by projecting the sphere on to the cylinder through, for example, rays from the centre of the globe. Mercator would have constructed his charts by careful measurement. In modern mathematical language, the vertical coordinate in the Mercator projection is proportional to the logarithm of the tangent of half the angle of latitude, a relationship later explored by James Harriot.

A Mercator projection greatly distorts areas away from the equator making Greenland for example, which is about the size of Mexico, look almost as large as Africa, although it does show more respect for the general *shape* of the island. There are many other map projections, each with their own uses: for example, on a so-called *gnomonic projection*, straight lines on the map do correspond to great circles.

Returning to the simple cylindrical projection, although it does not preserve *directions*, remarkably it does preserve *areas*, and this follows from an important result of Archimedes alluded to above. (It can be proved that there is no flat representation of the globe that preserves both areas and angles between lines at the same time.) The area of a cap of a sphere, that is the area between, let us say, the North Pole and some latitude line *l*, is equal to that which we get from projecting this cap horizontally on to the containing cylinder of the same diameter, the area of which is easy to calculate being equal to the circumference of the sphere times the height of the cap.[4] If we accept this fact, it follows that the slice of the sphere between any two latitude lines, l_1 and l_2, is also equal to the area of the slice projected on to the cylinder, as the area of the slice equals the *differ-*

---

3. They were not the first though, as Martin Behaim created a terrestial globe in 1492 that still survives and the first representation of the world as a globe is attributed to Crates of Mallus around 150 BC. Eratosthenes calculated the diameter of the earth in 230 BC through the difference in the sun's elevation at Syrene and Alexandria at the solstice.

4. This result is derived in the notes of the final chapter.

ence in the areas of the caps corresponding to l_1 and l_2, and so this difference equals the corresponding difference of the projections of the caps, which in turn is the same as the projection of the slice on to the cylinder (Fig. 4.3).

Next, take any such slice and imagine it cut into a large number of identical 'rectangles' whose sides are given by l_1, l_2, and some pair of longitude lines, L_1, L_2. Since these rectangles are identical, so must be their projections on to the cylinder. Since the sum of their areas is the area of the whole slice, each such rectangle, R, must have area equal to its projection, R', for if they differed, suppose R were less than R', for argument's sake, then the area of the slice would be less than that of its projection, which we know not to be the case. Finally,

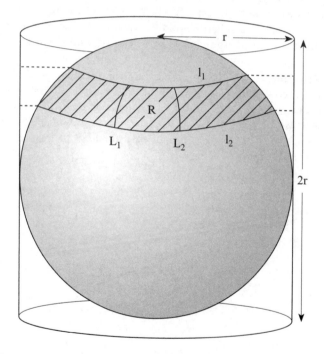

Fig. 4.3: Latitudinal slice projecting on to a cylinder

since an arbitrary area A on the sphere can be covered as accurately as we please by a collection of such little rectangles, it follows that A must equal its projected area A', for if these two quantities differed, a difference in the approximation of the areas of A and A' by rectangles would emerge for a sufficiently accurate approximation, and we now know this could not happen.

In particular, the surface area of a sphere of radius r, say, is equal to that of its containing cylinder. By opening the cylinder out flat into a rectangle we see that its area is given by the product of the height of this rectangle, $2r$, and its width, which is the circumference of the sphere, $2\pi r$. In this way Archimedes discovered that the area of the sphere is given by $4\pi r^2$. In his treatise *On the Sphere and Cylinder* he proves this and claims it as a new result. Once the surface area of the sphere is known, it is not difficult to see that the volume of the same sphere should be equal to its area multiplied by a factor of $r/3$. The argument is more easily appreciated if we first drop down one dimension and look at how the area of a circle relates to its circumference. Take a certain number, let us say n, equally spaced points around the edge of a circle of radius r. By joining each point to the centre, C, of the circle we form n identical triangles whose total area is an approximation of that of the circle itself. Each of these triangles has an area of half of its base b multiplied by its height h. (See Fig. 4.4.)

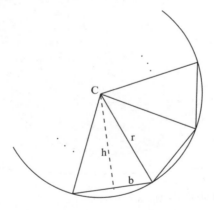

Fig. 4.4: Approximating a circle through triangles about its centre

The area of the polygon inside the circle formed by the bases of our triangles is $n(\frac{1}{2}bh)$. Now the area of the circle is the limiting value of this quantity as we use increasingly more, and so smaller, triangles in our approximation; since h then approaches r, the radius of the circle, while nb, the total measure of the outside of the polygon, approaches the length of the circle's circumference, $2\pi r$, we obtain the familiar result for the area of the circle: $\frac{1}{2} \times (2\pi r) \times r = \pi r^2$—that is, the ratio of the *area* of a circle of radius r to its *circumference* is $r/2$. As we shall see, when passing from the two-dimensional sphere (the circle) to the ordinary three-dimensional sphere the relationship between *volume* to *surface area* of the sphere is similar to that of *area* to *circumference* of the circle, except that the ratio between them changes from $r/2$ to $r/3$ when moving from two to three dimensions.[5]

In passing, it is worth noting how easy it now is to see what the area is of an ellipse. Imagine a circle of radius b and then stretch the circle horizontally away from its vertical line of symmetry by a factor of a/b: the outcome of this thought experiment is an ellipse of width $2a$ as in Figure 4.5. Since every horizontal strip of area (shaded in the diagram) has now been magnified by a factor of a/b, so has the overall area which must then equal $(\pi b^2)(\frac{a}{b}) = \pi ab$. By way of contrast, the problem of finding the *length* of the circumference of the same ellipse does not succumb to so simple an argument and presents a much more formidable problem leading to a new class of mathematical functions.

We can recover the formula for the volume of a sphere by the same type of argument based on cones instead of triangles, provided

---

5. The Chinese and Egyptians had useful approximations of π well over 3000 years ago. However it was Archimedes who established this basic relationship between the circumference of a circle and its area and moreover showed that the value of π lay between $3\frac{10}{71} = 3.1408$ and $3\frac{10}{70} = 3.1428$. In the 17th century the brilliant Japanese mathematician and engineer Matsunaga calculated π correct to 50 decimal places, which compares favourably to contemporaneous European efforts: Ceulan (1610) dedicated his life to finding the answer to 35 places while Sharp (1699) extended to 71 places using series techniques. Today the decimal expansion of π which has no pattern of any kind as far as we know, is known to millions of places. Mathematical historians are still left guessing however as to what method Matsunaga devised to accomplish his feat: see the web page http://www-groups.dcs.st-and.ac.uk

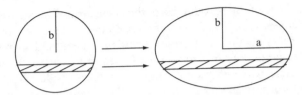

Fig. 4.5: Area of an ellipse

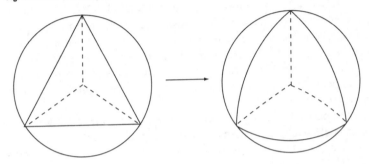

Fig. 4.6: Partitioning a sphere into triangles

that we know that while the area of a triangle is *half* the base times height, the volume of a cone is *one third* the base times height. We'll explain this fact shortly but, taking it for granted for the time being, we can finish the argument as follows.

Begin with a tetrahedron, that is, a pyramid consisting of four identical equilateral triangles. The lines from each corner of this pyramid to the face opposite meet at a point that we shall take as the centre of a sphere of radius *r*, say, so that the tetrahedron just sits within the sphere with its corners lying on its surface. By projecting the sides of the tetrahedron on to its containing sphere (by shining an imaginary light from its centre) we divide the whole surface of the sphere into four identical spherical triangles as in Figure 4.6.

Divide each triangle into three more triangles by joining each corner to a central point of the triangle. (It matters little what you take this central point to be—you could choose the centre of gravity or balance point of the triangle that we will talk about later in the chapter.) If we continue this process we shall partition the surface of the sphere into collections of ever smaller triangles. The triangles may not be identical, but the size of the largest will approach zero as we repeat the procedure again and again.

We now imagine joining the centre of the sphere to all corners of each triangle and in so doing we gain a better and better approximation of the volume of the sphere as a sum of triangular based pyramids. Now a pyramid is a kind of cone in so far as its area equals one third the product of its base and height. As the approximation improves, the common value of the *height* of these pyramids approaches *r*, just as did the height of the triangles in the case of the circle, while the sum of all the *areas* of the bases of the pyramids approaches that of the whole surface of the sphere, just as the sum of the *lengths* of the bases of the triangles in the circle case approached its circumference. For this reason we find that the volume of the sphere must be $r/3$ multiplied by the total surface area of the sphere, which, as Archimedes found, is $4\pi r^2$, four times the area of one of its great circles. Therefore the volume of our sphere is $\frac{4}{3}\pi r^3$ units cubed.[6]

A Look at Things Conical

There is nothing much easier to find than the volume of a box (the official name for it is *cuboid*) for it equals the area of the base times the height. The same is true however of any *prism* which consists of two identical bases, one lying directly above the other, joined by the surface formed by the perimeter of one base as it is moved directly on to the other (see Fig. 4.7). The volume of this object is once again the area of the shape multiplied by the height *h* of the prism. In fact there is no need for the second shape to lie directly above the first for this volume formula to hold. Imagine the prism on the left in Figure 4.7 to consist of a set of thin horizontal slices, each of which is also a prism of identical base but with very small thickness. The volume of the prism is then the sum of these smaller volumes. If each of these slices is moved a fixed distance to the right relative to

6. Any readers familiar with calculus might notice that, since the rate of increase of the volume of the sphere as a function of *r* equals its surface area, the derivative of volume should equal the sphere's area, as it does; in the same fashion, the rate of increase in the area of a circle with respect to *r* is equal to its circumference.

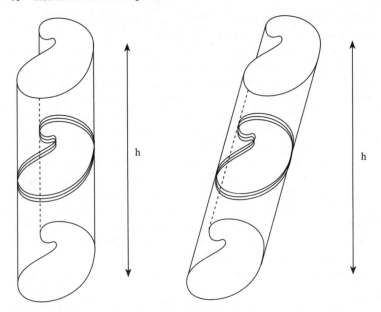

Fig. 4.7: Right and oblique prisms

the one it rests upon then the overall volume does not alter but the resulting three-dimensional object becomes much like a prism using the same bases which, although still parallel, do not lie one above the other. If we imagine these slices cut ever thinner, the volume of the object never changes but the object itself approaches that of an *oblique* prism rather than a *right* prism (Fig. 4.7). We are led to conclude that the volume of such a prism is equal to the area of its base times the *altitude* of the prism, which is the distance separating the planes of the two bases. This should not be a surprising result for you are likely to have seen this phenomenon in two dimensions before: the area of a parallelogram for example equals its base times its altitude.

A curious thing happens if we shrink the top face of the prism down to a point to create the corresponding cone. Take any point P of the top shape or *lamina* as we sometimes prefer to call it, and join it to a variable point Q on the perimeter of the bottom lamina (see Fig. 4.8). The corresponding *cone* is the surface that is generated by

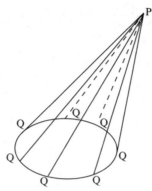

Fig. 4.8: Generating a cone

the line PQ, as Q traces out the perimeter of the lower lamina. Its volume V is always exactly one third that of the original prism in which it is embedded. That is, for any cone, $V = \frac{1}{3}bh$, where b is the area of its base and h is the height (or altitude) of the *apex* P of the cone above the plane of the base lamina. If we apply this general formula to an ordinary cone with a circular base we recover the familiar volume expression of $\frac{\pi}{3}r^2h$. To see why this fraction of one third emerges, let us begin by tackling the simplest type of cone, that being the pyramid.

By the slice-stacking argument that was employed previously for prisms, we see that the volume of a pyramid depends on just its base and height, and so two pyramids with the same base and height have equal volumes as the volume of each can be approximated to any degree of accuracy by the same set of stacked slices.

Let us first look at the case of a triangular pyramid $ABCD$. We can imagine this pyramid as part of a triangular prism also with the same base ABC and the same height h as shown in Figure 4.9. The prism itself can be cut into three triangular pyramids: $ABCD$, $DEFB$, and $ABED$. Although these pyramids are *not* identical, any two of them have the same base and height, although which face is taken to be the base depends on what pair of them we are comparing. (More detail of this argument is included in the final chapter: see Fig. 8.7.)

Granted that this is true, it follows that each of them, including our original triangular pyramid, has volume equal to one third of

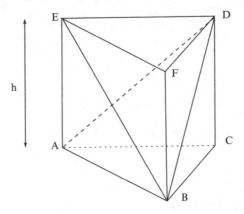

Fig. 4.9: Partitioning a prism into three pyramids

the volume of a prism with equal base and height, thus justifying our $\frac{1}{3}bh$ formula above. This rather deft partition argument can prove the point for triangular pyramids. For pyramids with arbitrary bases and more generally for arbitrary cones the argument proceeds by approximation through triangles somewhat reminiscent of the Archimedean argument for finding the area of a circle. (Again the final chapter notes have the details.)

Bodies in Balance

Much of Archimedes' mechanical intuition rested on the notion of balance. The following problem introduces the idea in a novel way through the question of cutting a cake in an equitable fashion. We have a rectangular cake as in Figure 4.10. The inner rectangle has a chocolate topping while the rest of the cake is vanilla (see Fig. 4.10).

How can you divide the cake into two helpings, using a single straight cut, so that both you and your friend enjoy equal amounts of chocolate and vanilla?

Your first impulse might be to cut the cake from the bottom left to the top right hand corner but a moment's thought will reveal that

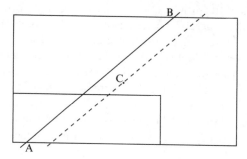

Fig. 4.10: Cutting the cake

will not divide the inner chocolate rectangle fairly as that cut does not match the diagonal of the smaller rectangle. Overall the solution must involve splitting the cake into two equal portions so let us ask first how this can be done. Any cut that goes through the middle, *C*, of the cake will at least achieve this much. The middle can easily be located as it is the point where the two diagonals of the outer rectangle meet. Cutting along *any* line through the centre will split the cake into two identical trapezium shapes so that both parts will be equal in size.

What is more, it is plain that we *must* cut through the centre, for if you draw a line *AB not* passing through the centre there is a line parallel to it that does meet the centre (again see the diagram). This parallel line splits the cake into halves so that our original line, which leaves a portion that is only part of one of these halves, would fail to divide the cake evenly.

We therefore must choose a line through the centre of the cake but which, if any, divides both toppings evenly? By the same reasoning as before, any cut that splits the chocolate topping must pass through the centre of the inner chocolate rectangle and so there lies the one and only solution to our problem—we find the centres of the outer and inner rectangles and these two points define the line of the cut. That cut splits the chocolate rectangle in half, and also splits the cake in half, and so must share out the remaining vanilla topping evenly as well.

The solution was, in the end, simple enough but we are entitled to ask to what extent the reasoning behind the solution depended on the specifics of the problem. For instance, was it important that the shapes involved were rectangles? Not really: the only fact that we used about the rectangles was that each had a *centre*.[7] If we had a circular or even an elliptical cake, or if the chocolate region itself was a circle the problem would have the same solution: we could divide the cake fairly only by joining the centres of the two figures involved. Symmetric figures such as circles, ellipses, and rectangles have an obvious centre but what of figures that have not? The simplest plane figure that generally is not symmetric is the triangle so we begin there.

The Middle of a Triangle

Returning to the books of Euclid (*c.* 300 BC), we see that the ancients recognized four different candidates for the title of centre of a triangle. To explain them let us look at the triangle △*ABC* in Figure 4.11 where we have labelled the sides opposite the vertices *A*, *B* and *C* by the letters *a*, *b*, and *c* respectively.

If we draw the line that *bisects* the angle at *A*, (dotted in Fig. 4.11), that is to say, cuts the angle into two equal halves, then all points on that line are an equal distance from each of the sides *b* and *c*. Similarly, the bisector of the angle at *B* is equidistant from the sides *a* and *c*. Let us call the point where these bisectors meet by the

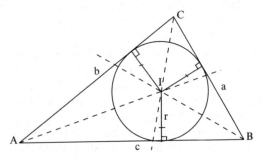

Fig. 4.11: Triangle, incentre, and incircle

---

7. All the same, it can be proved that there is always a single straight line that slices each member of *any* pair of regions precisely in two.

name I, and write r for the distance from I to c. Now since I is equidistant from b and c, the distance from I to b is also r, and since I is also equidistant from a and c its distance to a is r as well. What is more, since I is then equidistant from a and b, it lies on the bisector of the angle at C also. In summary, the three angle bisectors of the triangle meet at a common point I that is the same distance r from each of the three sides of the triangle. A circle drawn with centre I and radius r will just touch the three sides. This circle is sometimes called the triangle's *incircle* (it is the largest that fits inside) and for that reason the intersection of the three angle bisectors of the triangle is called its *incentre*: it is the point equidistant from all three *sides* of the triangle.

If instead of bisecting the *angles* of the triangle we bisect the *sides*, something similar happens (see Fig. 4.12). The perpendicular bisector of side a consists of all points equidistant from B and C and that of side b is the line of all points that are equally far from C and from A. Where these two lines cross we have a point O that is equidistant from all three *vertices* of the triangle so that a circle centred at O passing through one of the vertices of the triangle in fact passes through all three. This circle is called the *circumcircle*, it is the *smallest* that contains our triangle and its centre, O, is known as

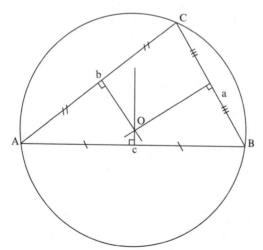

Fig. 4.12: Triangle, circumcentre, and circumcircle

the *circumcentre*, although this centre will lie outside of the triangle if it contains an obtuse angle. Again, since O is equidistant from A and B it must also lie on the perpendicular bisector of the third side c and so the three perpendicular bisectors of the sides of a triangle are *concurrent*, meaning that they meet at one point, and that point is the same distance from the three *vertices* of the triangle, as its corners are usually called.

Neither of these two rival centres of the triangle however is its balance point: that is to say the triangle, or rather a cardboard replica of it, would *not* be balanced if you tried to support it from below on your fingertip at either the incentre or the circumcentre. The balance point of a figure is called its *centroid* or *centre of gravity*. The most straightforward way to locate the centroid, G, of a triangle is as the common intersection point of its three *medians* which are the lines that go from a vertex of the triangle to the midpoint of the side opposite (see Fig. 4.13).

It is not difficult to convince yourself that a triangle would be in balance when placed on an edge lying along the line of a median. Imagine the body of the triangle as made up of thin lines, like the line DE shown in Figure 4.14, parallel to the side AC. What ensures the balance is the fact that the median BM also bisects each parallel line segment DE at its midpoint N. Accepting this allows us to infer that the centroid of the triangle lies on each of the three medians which must therefore be concurrent.

To see that N truly is the midpoint of DE we resort to comparison of similar triangles, that is, triangles that are the same shape although perhaps of different size. Each side of one triangle of such a

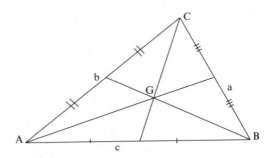

Fig. 4.13: The centre of gravity of a triangle

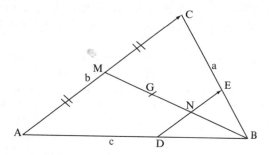

Fig. 4.14: Balance along a median of a triangle

similar pair is a fixed multiple of the size of the corresponding side of the other. Two triangles are similar exactly when they share the same set of three angles. For example, the triangle $\triangle BAC$ in Figure 4.14 is, for this reason, similar to the triangle $\triangle BDE$ making the following pair of ratios the same:

$$\frac{DE}{AC} = \frac{BD}{BA}.$$

In the same way triangle $\triangle BDN$ is just a smaller version of $\triangle BAM$ and so

$$\frac{BD}{BA} = \frac{DN}{AM}.$$

Combining these two equalities gives us

$$\frac{DE}{AC} = \frac{DN}{AM}.$$

Finally, since AM equals half of AC, this third equation yields that DN is half of DE, and, of course, NE is the other half. We see therefore that DE is exactly balanced on the median line BM.

Another symmetry of the medians is that they trisect one another. In other words, for each median, the centroid G of the triangle lies exactly one third away along the median when travelling from the midpoint of a side back to the opposite vertex: thus, in Figure 4.14, MG is one third the distance from M to B. The lines parallel to each side of a triangle and of one third the altitude meet at the centre of gravity of the triangle and so, considered as a lamina, the triangle would be balanced when resting on any one of them. We shall have more to say about this shortly.

The fourth centre known to the ancients is that which results from the intersection of the three altitude lines of a triangle, an *altitude line* being a line perpendicular to one side through the vertex opposite. Once again it can be shown that these three lines are coincident although this point, known as the *orthocentre* will also lie outside the triangle if it is obtuse.[8]

And there are others: a fifth centre F was discovered by Fermat in the 17th century: to find the *Fermat point* of a triangle we place equilateral triangles on each of the sides of a given triangle T and join each vertex of T to the far vertex of the equilateral triangle on the side opposite. The three lines drawn do turn out to be concurrent and this new centre, F, has the property that the three angles *AFB*, *BFC*, and *CFA* are all equal. What is more, F is the solution of a certain optimization problem in that the sum of the distances of F to each of the three vertices of T is as small as possible if T has all angles less than $120°$, a fact proved by the 17th-century Italian mathematician Torricelli.[9]

Balance Points

Moving on from simple triangles we ask the question, 'where lies the very heart of England?' The claim is made that the north-east Warwickshire town of Nuneaton, originally the home of the writer Mary Ann Evans (better known by her pen name of George Eliot), is the closest large town to the centre which has been given in a letter by the Royal Geographical Society at latitude $52° 33'$ north and longitude $1° 28'$ west, placing it on the Watling St, 4 miles ESE of Atherstone, close to a railway bridge, between the villages of High-

---

8. The centroid, circumcentre, and orthocentre lie in line, a fact that we may have expected to find in Euclid but went unnoticed until proved in the 18th century by the outstanding Swiss mathematician Leonhard Euler: in his honour this line of centres is known as the *Euler Line*.

9. The hunt for new centres goes on, there are many new ones to be seen on the web page: http://cedar.evansville.edu/ck6/tcenters/

am-on-the-Hill and Caldecote. If this were true, it would mean that a template of England would be in balance if placed to rest on the point representing the location marked by these coordinates.

This can be put to the test as the centre of gravity can be easily discovered for any lamina in a simple and practical way. Take any shape, as pictured in Figure 4.15. A shape like this that is regarded as flat and of negligible thickness is referred to as a *lamina*; we shall assume that our laminas are of *uniform density*—that is to say, consist of a layer of identical point particles spread evenly throughout the object.[10] The way one might find our balance point in practice would be first to take a fine edge, such as a rule, and move the shape about on top of the rule until it was neither inclined to tip one way nor the other—in other words, balance the shape on the edge of the ruler. Any balance point of the shape must lie along this edge. If we rotate the shape and then find another such balance line then the point where the two lines meet will be the centre of gravity—if the shape were supported from below at that point it would not be inclined to topple in any direction. Any number of such lines of balance could be found and they will meet at the centroid, that is to say they will share the centre of gravity as their common point, as the balance point must lie on every line of this kind. An alternative approach would be to find a pair of these lines of balance using a plumb line as shown in Figure 4.15.

Fix the shape at one point in such a way that it is free to swing and so comes to rest in a natural attitude of balance. The

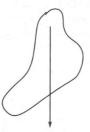

Fig. 4.15: Plumb line from point of suspension of a lamina

10. The centroid is the centre of mass of a uniform lamina: if the mass is not evenly distributed then the centre of mass may lie elsewhere other than at the centroid.

shape will then be balanced about the vertical line through the point and so a plumb line from that point, marked by a thin weighted string, runs down the line of balance for your shape. Any number of such lines could be constructed and will coincide in the same point. Applying this idea to a map of England we obtain the pictures in Figure 4.16.

The same method will work with a three-dimensional object. Plumb lines dropped from suspension points all coincide in the centre of gravity of the object, which is the unique point that lies below any point by which the object is suspended.

The centre of gravity is an important concept in the calculation of the mechanical behaviour of an object as much of the object's behaviour is identical to that of a point object of the same mass concentrated at the centre of gravity. Study of this relatively simple (although imaginary) object can then be used to describe the behaviour of the real thing. It can even be used to calculate some purely mathematical quantities such as the volumes of solid objects including doughnuts and cones, as will be explained later in the chapter.

Returning to our original problem concerning the rectangular cake (Fig. 4.10), we see that the centre of the rectangle in the usual geometrical sense (the intersection of its diagonals) also coincides with its centre of gravity. However a line through the centre of gravity of a lamina does not necessarily leave equal mass on either side (and so, not equal areas either). For example, the balance point of the lop-sided dumb-bell shape in Figure 4.17 will be somewhere along the thin bar joining the two circles leaving the greater area on the right. If you imagine the dumb-bells pivoted on a fulcrum at the point G it would not tip either way. What is in balance at the centre

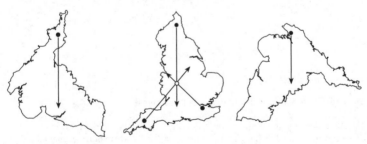

Fig. 4.16: Centre of England

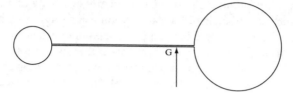

Fig. 4.17: Offset centre of gravity

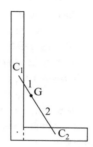

Fig. 4.18: Centre of gravity of an L-shape

of gravity is the *torque* or *moment*—the tendency to turn about the point. This principle can be seen in action in construction cranes where the overhead gantry is very lop-sided but is kept in balance by weighting of the shorter section.

The contribution of each atom of the lamina to the torque depends both on the mass of the atom and its *distance* from the line of the action—in fact it is calculated by multiplying these two quantities together. It is this overall turning effect which is in balance at the centre of gravity rather than the areas either side of any line passing through it. In the case of figures symmetrical about such a line though, like our cake, the areas either side will also be equal.

The centre of gravity may lie outside the object itself. For example the centroid of a hoop is at the centre of the circle that it surrounds. Even shapes which do not contain holes can have external centroids—take for instance a cut-out of the letter L (see Fig. 4.18). The centroid will lie inside what in the first chapter we called the *convex hull* of the shape, which is obtained by imagining the object placed inside a loop of string that is then pulled tight. This is because a convex shape always contains the line running between any two of its points. In particular the centre of gravity of a convex shape such as a triangle or regular polygon will always lie within the figure so that such shapes have an internal point of balance.

Returning to our letter L, we may ask if there is a simple method by which we could calculate its centre of gravity, for nailing copies of the letter to walls and dropping plumb lines looks unnecessarily awkward for so simple a shape. Indeed we can and the calculation affords us an example of a general principle that we shall look to again a little later. If we have a complex lamina, one that can be regarded as consisting of two or more non-overlapping parts, we may find its centroid as follows. First, find the centroid of each of the parts. Then treat the object as a collection of particles, one at each minor centroid with mass proportional to the weight of the corresponding part. The centre of gravity of this collection of imaginary particles is the same as that of the original figure.

Applying this to our letter L, we may choose to regard the lamina as consisting of two connected rectangles as pictured in Figure 4.18. To be definite, let us suppose that each rectangle has equal thickness but that the stem of the letter is twice as long as the base (and so has twice the mass). The centroid of each rectangle is then just the centre of each. We now think of the lamina as equivalent to two point particles, one at the centre of each rectangle, with that representing the longer rectangle twice the weight of the other. The centre of mass of this particle pair then lies on the line joining them—but not in the middle. To keep the imaginary see-saw in balance, the balance point would have to be twice as close to the *heavier* mass as to the lighter one. That is to say, the centroid, G, of the letter L is on the line joining the centres of the two rectangles with the distance $C_2 G$ equal to twice the distance $C_1 G$. Despite lying outside of the body, the centre of mass still has physical meaning: if the *L* were tossed into the air like a boomerang it would rotate about its centroid as it flew so the L still 'knows' where its centre of mass is.

We could apply this principle to three or more particles without real difficulty. If the location of the centroid of the set of particles was not obvious through symmetry considerations, we could work on one pair of particles at a time, first replacing one pair by a single point representing their combined mass lying at the centre of mass of the pair, and proceeding in that way, reducing the number of particles by one at each stage of the calculation until just one point mass remained representing the entire mass of the lamina concentrated at its centre of gravity.

Why should this work? Imagine any shape you wish and slice it mentally into two parts, P_1 and P_2, with a line (which need not be a straight cut). Let G_1 and G_2 be the respective centroids of the parts. Since the parts are no longer attached to one another (although imagine that they are still touching), P_1 will happily stay balanced sitting on G_1 and likewise so will P_2 on G_2. It follows that the original shape would be balanced if left sitting on the line joining G_1 to G_2 and so its centroid, G, lies somewhere along this line. Since a point particle placed at the centroid of each part and of equal mass has the same turning effect as the part itself, the centroid, G, will be at the centre of gravity of these imaginary masses.

As a general observation, imagining an object to consist of individual particles (atoms it you like) can sometimes explain physical behaviour that might otherwise seem strange. For example, take Galileo's famous observation that two objects fall under gravity in exactly the same way. For instance, a basketball and a lead cannon ball, when dropped from the Leaning Tower of Pisa, should strike the ground at the same time. Well perhaps not, but that is only because air friction acts on the two objects differently. If the experiment were done on the moon (no air) they would strike the ground at the same time. If you are old enough, you may remember this being demonstrated live on television in 1970 by an Apollo astronaut who dropped a hammer and a feather on the moon from the same height and we watched as they fell to the ground together. (Slowly, because of the moon's lesser gravity, and silently, as there was no air to transmit sound, but they did land simultaneously as Galileo predicted they would.)

There is little mystery however if you regard each object as made up of atoms, separated from one another by relatively large distances. When released, each atom falls under gravity and it does not 'know' whether it is part of a cannon ball, a basketball, or anything else. They all simply fall together with the same constant acceleration due to the constant force of gravity acting on each one.

Finally, let us return to the claim we made earlier in the chapter that the centroid of a triangle cuts each median in the ratio of two to one. Instead of recounting the geometric proof of Euclid we give a mechanical argument more in tune with the theme of the chapter. We begin with the simplest case of an equilateral triangle and mark the points that divide its sides into thirds. These points form a regular hexagon whose centroid G is, by symmetry, at its centre which can be easily located as it lies at the intersection of the diagonals of the hexagon (see Fig. 4.19). Imagine the hexagon balanced on its centre and mentally adjoin the three smaller equilateral triangles to the outside of the hexagon to reform the original triangle.

Since these triangles are symmetrically placed about G, the new figure would remain in balance and so G is the centroid of the triangle. (As you would expect, all conceivable types of centre of an equilateral triangle coincide.) The balance line L of the triangle parallel to the base BC must pass through G and so is evidently at an altitude which is one third that of the triangle.

So much for equilateral triangles. We can now make our equilateral triangle into an arbitrary isosceles (two sides equal) triangle by suitable stretching. Take each point of the triangle in Figure 4.19 and move it *perpendicularly* to the line L by increasing its distance from L by a fixed factor c. This will give a new triangle in which both the sides AB and AC will be equal while the length of BC will not change—the sides AB and AC will be longer than BC if c is greater than 1 but shorter if c is less than 1 (see Fig. 4.20 where $c = 1.5$).

Consider how the moment of the triangle about the line L has changed. (Note that L is unaffected by the stretching.) An atom whose distance from L was originally x is now at a distance cx from

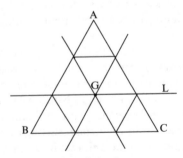

Fig. 4.19: Centre of gravity of an equilateral triangle

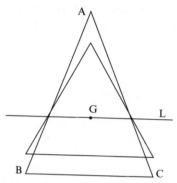

Fig. 4.20: The centre of gravity is invariant when the equilateral triangle is stretched perpendicularly from *L*

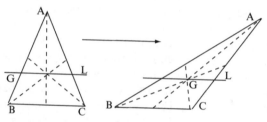

Fig. 4.21: Centre of gravity of an arbitrary triangle

our balance line. This will increase the overall moment on *both* sides of *L* by a factor *c*, so that the new triangle will still remain in balance about the line *L*, which evidently remains parallel to *BC* and at a distance of one third the altitude of the new triangle.

Finally we can turn our isosceles triangle into a *scalene* triangle (one of arbitrary shape) by a *shearing action* about the line *L*. We move each point of the isosceles triangle $\triangle ABC$ a distance which is proportional to its distance from the line *L*, the points above the line moving right while those below moving left. Once again, no point on the line *L* moves at all under this transformation and the sides of the triangle will remain straight (Fig. 4.21). The distance of each atom from the line *L* does not change as every point is moving parallel to *L* which still has the same relative altitude as it did before. Hence the triangle remains in balance when resting on a line parallel to the given side *BC* and with altitude one third that of the triangle. The meeting point of these three lines, one parallel to each side, will therefore mark the centroid *G*.

Centroids of Some Solids

One mathematical application of the notion of centroid is through a pair of remarkable facts known as the *Theorems of Pappus*. Pappus of Alexandria flourished early in the 4th century of our era. His eight books, forming what he described as *The Collection*, is a summary of much of the best of a millennium of Greek mathematics and also contains the new discoveries of his own age. It represents the swan song of Hellenistic mathematics for it marks the end of the classical period of mathematical development.[11] For the next thousand years mathematical progress, while not insignificant, was to be very sporadic and lack overall direction. The very idea that scientists and mathematicians should always be looking to break fresh ground, while alive in the minds of stubborn individuals, was never sustained and, in Western Europe especially, the notion of mathematical progress seems to have vanished entirely. Fortunately some real progress was made in the Muslim world during the medieval period from which Europe benefited greatly during the Renaissance and beyond. The long mathematical winter was endured but not without some real losses. Not even all of Pappus' *Collection* has come down to us intact, although his own best results on solids and surfaces of revolution survived and will be explained here.

Let us begin by introducing an example through which we can illustrate the theorems. A *torus* is the doughnut-like shape that results when a circle is rotated around an axis line outside the circle but within its plane—the circle will sweep out a kind of tunnel which eventually joins back with itself as pictured in Figure 4.22.

11. The martyred Roman scholar Boethius (*c.*475–524) wrote a work consisting of some Euclidean geometry and arithmetic. This relatively minor effort was important as it survived in monastic schools for centuries where it was regarded as the pinnacle of mathematical achievement and thereby a link was maintained between Christian Europe and the Golden Age of Alexandria.

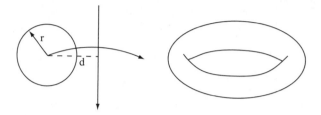

Fig. 4.22: Generation of a torus

Pappus' First Theorem tell us how to calculate the volume of such a *solid of revolution*. More importantly, as Pappus proudly proclaimed, the theorem not only works for circles but for any shape at all: the volume created is equal to the area of the shape multiplied by the distance travelled by its centroid. Thus we see the centre of gravity entering into the solution of a question which may be thought to be not one of mechanics but of a purely mathematical character. A torus is specified by the radius of its generating circle, r, and the distance, d, of the centre of this circle from the axis of revolution. The distance travelled by the centre of the circle is therefore $2\pi d$ and its area is πr^2. Pappus then tells us that the volume V of a torus is given by

$$V = (2\pi d)(\pi r^2) = 2\pi^2 d r^2.$$

What Pappus had done was to reduce the problem of finding such a volume of revolution to that of locating the centroid. Since we have gone to some length to solve the centroid problem for triangles let us apply Pappus there. Rotating a right-angled triangle as shown in Figure 4.23 about one of its shorter sides results in a cone.

As we now know from our work on triangles, the distance of the centroid of this triangle from the axis of rotation is $r/3$, and the area of the triangle is half the length of the base times the height, $rh/2$. Pappus tells us that the volume, V, of our cone is therefore given by the product

$$V = (2\pi \tfrac{r}{3})(\tfrac{rh}{2}) = \tfrac{\pi}{3} r^2 h.$$

Pappus' Second Theorem is to surface areas what his First Theorem is to volumes. The surface area of a solid of revolution equals the length of the *perimeter* of the shape multiplied by the distance

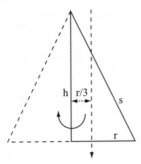

Fig. 4.23: Generating a cone from a triangle

travelled by the centroid *of the perimeter curve* (which may be a point different from the centroid of the shape itself).

A simple example of this is provided by the torus, as here the centroid of the perimeter circle coincides with the centre of gravity of the corresponding disc. Once again, the distance travelled by the centre of the circle in sweeping out the torus is $2\pi d$, while the perimeter of the circle is its circumference with length $2\pi r$, and so Pappus tells us that the surface area, S, of the torus is given by:

$$S = (2\pi d)(2\pi r) = 4\pi^2 dr.$$

Similarly, the surface area of the cone is easily expressed in terms of its *slant height*, which is the distance s down the slope from the apex to its base. This is because the surface is generated by rotating the slanting edge about what then becomes the axis of the cone. The centroid of this edge is obviously half way along, placing it at a distance of $r/2$ from the axis of rotation. The distance travelled by the centroid is thus $2\pi(r/2) = \pi r$, giving the surface area, S, of a cone to be πrs. If we wish, we can express s in terms of the height and radius of the cone for, by Pythagoras' Theorem, $s^2 = r^2 + h^2$ and therefore by taking square roots we can write the surface area S as

$$S = \pi r \sqrt{r^2 + h^2}.$$

The difficulty in applying the Pappus theorems is in finding the centroid of the given figure, which can be hard if we cannot see the answer immediately through symmetry. For example, where lies the centre of gravity of a semicircle? If we knew we could then find the surface area and volume of a sphere through use of Pappus. We can, however, take the opposite tack. Since Archimedes has taught us about the surface of a sphere we can then use Pappus in reverse to calculate our centre of gravity and we shall return to this in a moment.

Archimedes was a genius and, we now realize, a very complete mathematician, for on the one hand he insisted on the highest standard of proof in his work, while on the other he appreciated the value of intuitive and physical approaches to problems as a path to discovery. This more complete insight into the thinking of Archimedes did not emerge until 1906 when one of his manuscripts, *The Method*, which had been lost for almost one thousand years, was rediscovered by the Danish scholar Heiberg on a parchment in Constantinople. The parchment had been used to copy *The Method* sometime during the 10th century only to be scrubbed off in the 13th century in order to record a certain collection of Eastern Orthodox prayers and liturgies. Luckily the washing process was not thorough enough to totally destroy the original and through careful study and photographic techniques the scholar reconstructed most of the original work of Archimedes.[12] It was doubly fortunate as not only does this palimpsest still represent the only source of this precious manuscript but *The Method* of Archimedes is different from all his other known publications. In it he explains how his primary mathematical discoveries were made through heavy use of the intuitive idea of moments being in balance around a fulcrum. He even goes on to point out the shortcomings in this approach in that it assumes, for example, that an area can be regarded as a sum of line segments which is an idea that does not stand up to scrutiny if pursued but which can sometimes lead to correct answers to otherwise difficult questions, as we ourselves saw earlier when discussing balancing a line in a triangle about a median.

---

12. Heiberg's discovery made the front page of the *New York Times* on 16 July 1907 yet the manuscript itself mysteriously disappeared from public view for decades only to turn up for auction at Christie's in New York in 1998. It is currently being cared for and analysed by the Walters Art Museum in Baltimore. It seems now that the treatise was written as a letter to Eratosthenes and the ironic tone of the introduction suggests that Archimedes knew he was a long way ahead of his time and that Eratosthenes would have his hands full coping with it!

In view of Archimedes intellect, his understanding of geometry, and his heavy use of mechanical intuition as a source of mathematical discoveries, the theorems of Pappus are just the kind of results that you would have expected from him. Archimedes certainly knew more than is directly disclosed to us in his existing works, as is evidenced by medieval Arabic scholars who, having access to documents that we do not, tell us about other results known at that period such as the Heron Formula for the area of a triangle in terms of the lengths of its three sides. Six hundred years after Archimedes, Pappus also would not have had available all the best work that had gone before. Even Pappus lost credit for the centroid theorems for a time as they were rediscovered by the early 17th-century Swiss mathematician Paul Guldin and are sometimes called by his name.

Tricky volume and area formulae nowadays are usually considered to be the business of integral calculus texts. Yet as we have seen, they can be deduced using more elementary algebraic and geometric techniques. Indeed many of the standard problems of this kind studied by calculus students had been solved more than adequately by Archimedes and his fellow Greek mathematicians through use of their own techniques that now are largely neglected except by mathematical historians. Further examples include that of a wedge cut from right circular cylinder, the volume of two intersecting cylindrical pipes, areas involving spirals, and the area under part of a parabola (the curve of a projectile under gravity). Lacking the techniques of integral calculus that allow calculation involving infinite processes, Archimedes would nominate a candidate for the answer to his question (the answer itself no doubt having first been discovered using the techniques allied to those of his *Method*), and then, using what were already in his own day time honoured methods to be found in Euclid and elsewhere,[13] would show through a finite calculation that any different answer must be wrong. This rather unnatural approach, an inspired guess for which no explanation was offered, followed by a kind of mathematical method of elimination, was not to be improved upon for 1900 years till the days of Isaac Newton.

---

13. *The Method of Exhaustion*, probably invented by Eudoxes to circumvent the so-called Paradoxes of Zeno concerning infinite divisibility, is described in Book 12 of *The Elements*.

To close this chapter we look at how Archimedes and Pappus come together to allow us to find, in a trice, the centre of gravity of a semicircular disc of radius r, although the answer is not something that might be guessed. By symmetry, the centroid of the semicircular disc lies on the line of symmetry through the centre but what still remains to be found is the distance d of the centroid from the centre of the circle (see Fig. 4.24).

Rotating the semicircle about its diameter gives a sphere that Archimedes tells us has volume $\frac{4}{3}\pi r^3$. On the other hand, Pappus tells us that this volume also equals the area of the semicircle, $\frac{1}{2}\pi r^2$, multiplied by the distance the centroid travels as the rotating disc sweeps out the sphere, which is $2\pi d$, the length of the circumference of the circle of radius d. Equating these two expressions for the volume of the sphere gives us

$$\frac{4}{3}\pi r^3 = (2\pi d)(\tfrac{1}{2}\pi r^2) \text{ and so } d = \frac{4r}{3\pi}\, (\, \approx 0.42r).$$

As an application of the surface theorem of Pappus we can find the centre of gravity of a wire semicircle of radius r. Again let d be the distance from the diameter of the wire to the centroid. This time we equate the surface area of the sphere generated by rotating the wire about its diameter $(4\pi r^2)$ to the distance travelled by the centroid $(2\pi d)$ multiplied by the length of the semicircle (πr) to give us

$$4\pi r^2 = 2\pi^2 rd \text{ and so } d = \frac{2r}{\pi}\, (\, \approx 0.64r).$$

The answer to the first question has turned out to be two-thirds that of the second and this is due to the fact that the area of the semicircle can be approximated as accurately as we please by a series of identical isosceles triangles based around its rim with common vertex at the centre of the disc; the centre of gravity of each of these triangles is two-thirds their distance from the centre so that the

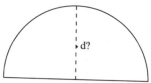

Fig. 4.24: Centroid of a half-disc

centroid of the disc is at the same point as the centroid of a semicircular wire with the same centre but with two-thirds the radius.

Archimedes was the outstanding figure of the Alexandrian age, but the centuries that lie between him and Pappus produced some particularly important discoveries, although again the details are not always available. Eratosthenes estimated the diameter of the Earth in 230 BC and also left as a legacy his 'sieve' method for the generation of all prime numbers. Diophantus, an important but extremely obscure figure, probably lived around the year AD 250[14] and his treatise on number theory inspired Pierre de Fermat to devise his celebrated Last Theorem that was finally proved in the 1990s by Sir Andrew Wiles. Heron of Alexandria was a leading light of the first century of the Christian era, while Apollonius of Perga produced the definitive account of what are known as conic sections, which enter our story in the next chapter.

---

14. We do happen to know how long he lived as someone in the 5th or 6th century left behind this riddle: Diophantus lived one-sixth of his life as a boy, one-twelfth as a young man, and he married after a further one-seventh part; five years after he wed, his wife produced a son who lived only to half the age of his father and died four years before him. Formulating all this as an equation you can deduce that he lived to 84. (Details in the final chapter.)

5 ○ Reflections and Curves

Much of this chapter has to do with reflections from curved surfaces but to whet your appetite, let us begin with a question about ordinary flat mirrors. When inspecting your appearance in a mirror that is not quite long enough to show your full figure, a natural impulse is to step back a pace or two in the hope that this will allow the mirror to accommodate. It doesn't work, so why do we instinctively expect that it will? It may be because, while looking at three-quarters of your body, you can see in the background large objects, perhaps other people, in full view. Although *you* can see the full figure of the person in the background, they however cannot and it will do you no good to step back to where they are. Let us look at the geometry in order to see why this is so.

How long must a mirror be in order for you to see your full reflection?

In Figure 5.1 the line segment on the left represents you, that on the right the mirror, and *E* is your eye level. Let the distance above your eye be denoted by *a* and that below your eyes be *b* so that your

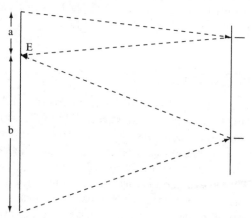

Fig. 5.1: Full-length reflection

own height h is $a + b$. In order to be able to see the top of your head in the mirror, light travelling from there has to be able to strike the mirror and reach your eye. Since light rays reflect at the same angle with which they hit a surface, by symmetry the light ray in question must strike the mirror at a distance of $a/2$ above your eye level. Similarly, the mirror must stretch at least a distance $b/2$ below your eye level, for then you will be able to see your feet. We see that the mirror can just manage to produce a full reflection that is visible to *you* as long as its height is at least $\frac{a}{2} + \frac{b}{2} = \frac{h}{2}$. In other words you can see your full reflection in a mirror provided that the mirror is at least half your height: a 3 foot mirror will be good enough for most people.

Note also that the distance you place between yourself and the mirror did not enter into the reckoning, so there is no use in stepping forward or back. The vertical position of the mirror however makes some difference—if it is of the minimum satisfactory length, then the segment that lies above eye level has to just match your own height above the mirror top. This means in practice that you should mount the tops of so-called full-length mirrors about 4 or 5 inches above your eyebrows if you want to see how your hat goes with your shoes.

An interesting variant of this problem that you might like to think about is where you have two people, a shorter one of height h and a taller of height H say, standing side by side before a mirror. How long a mirror do they need so that each of them can see the reflection of both themselves and their partner?

The Heron Problem

A simple reflection problem that has had surprising echoes down the ages is due to the Alexandrian scientist Heron. Heron may have been Egyptian rather than Greek but again the existing evidence is scanty and places him only somewhere in the time line 150 BC to AD 250. He is known to us mainly through his surviving works, among which are *The Dioptra* and his *Metrica*. Like Archimedes he worked in engineering and mathematics but unlike Archimedes was equally keen on practical inventions, said to include the first working steam engine, a fire engine, a primitive thermometer, and various toys and gadgets for opening temple doors and the like. In all this restless energy we are reminded of Leonardo Da Vinci for we might suppose that not all of his brilliant ideas were realized. On the mathematical side he devised the iteration that is still sometimes taught in schools to calculate square roots and the formula for the area of a triangle in terms of its three sides bears his name:

$$A = \sqrt{s(s-a)(s-b)(s-c)},$$

where s is half of the triangle's perimeter. It is said that Archimedes knew and had a proof of this but it continues to be known as the Heron Formula as Heron's clever proof is the oldest that has come down to us through a complete copy of the *Metrica* that was unearthed in 1896 in an Arabic manuscript which itself dates to around the year 1100. The Heron Formula is Proposition 1.8.

The problem I draw your attention to here though comes from Heron's *Catoptrica*, the subject of which is mirrors and the behaviour of light. The Heron Problem is portrayed in Figure 5.2.

Find the point P on the line L that minimizes the *sum* of the distances $F_1 P + P F_2$ of the two fixed points F_1 and F_2 on the same side of the line as L.

This problem is sometimes given as an optimization problem in differential calculus but Heron himself quickly arrived at the

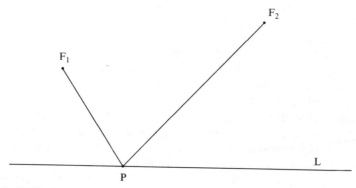

Fig. 5.2: The Heron Problem

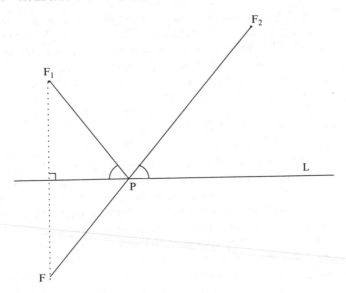

Fig. 5.3: The Heron Problem solved

solution through use of reflections as follows. Let P be any point on L and reflect the line F_1P in the line L to give the line FP, as shown, of the same length as F_1P. We see that the length of the path F_1PF_2 is equal to that of the path FPF_2 and so we may choose P to minimize the former by selecting the point P to minimize the latter. But this is easy for to make FPF_2 as small as possible we should choose P so that the three points F, P, and F_2 lie in line. In this way we have discovered our optimum point P—it is the point where L meets the line FF_2, where F is the reflection of F_1 in L.

Heron went on to note that this optimum point is characterized by the special property that the angles that the lines F_1P and F_2P make with the line L are equal. He then inferred the Law of Reflection of light that the angle of the incident ray equals the angle of reflection from the assumption that light is naturally efficient and would take the shortest path possible from F_1 to F_2 via the reflecting line L. 'Nature never acts in vain', as he elegantly put it. By taking the shortest route the light ray would be travelling from one point to the other in the least possible time and expressed in this way the principle is often known as Fermat's Principle of Least Time and is seen to apply to light refracting between transparent media as well, such as a light ray travelling from fish to eye from a goldfish bowl.

This style of problem where a point needs to be found that minimizes a certain distance occurs elsewhere in geometry and can sometimes be tackled in the same way: we begin with an arbitrary point P and show that the distance associated with P is equal to that of a certain bent line—the length of that line is minimized by making it straight and that leads to the best choice for P. This is how the above solution to the Heron Problem worked and an argument along these lines succeeds in showing that the Fermat centre F of a triangle (see p. 86) is the point that minimizes the total distance to the three vertices.[1]

The first surprise the Heron Problem has in store for us is that it tells us something important about ellipses. Recall that an ellipse is the set of points P that have the property that the sum of the distances $F_1P + PF_2$ is a fixed number, (see p. 28). Consider such an ellipse and let L be a tangent line to the ellipse at a point P on the curve as shown in Figure 5.4.

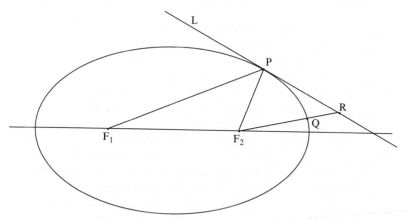

Fig. 5.4: Reflection Property of an ellipse

---

1. This solution, due to J. E. Hofmann, can be found in *Introduction to Geometry* by H. M. Coxeter.

For any other point R on the line L the sum of the distances to the foci exceeds that of the corresponding sum for P, a fact that we shall explain more carefully in a moment. Granted that this is so however, it follows that P is the solution of the Heron Problem for the points F_1, F_2, and the line L. It then follows, as Heron observed, that the angles that the lines F_1P and F_2P make with L are equal and so it follows from the Law of Reflection that a light ray from one focus of the ellipse is reflected by the ellipse back to the second focus.

Having seen the remarkable conclusion to this argument, let us return to the claim that it rests upon by considering another point R on the tangent line L of our ellipse, as in Figure 5.4. The ellipse is a closed curve that is convex everywhere, and at each point it has a tangent that remains outside of the figure everywhere except at the contact point P. It follows that the line from F_2 to R meets the ellipse at some point Q as shown. Now since Q lies on the ellipse the sum of its distances to the two foci is the same as that of P which justifies the first equality in the next statement, while the others are easily seen from the diagram.

$$F_1P + PF_2 = F_1Q + QF_2 < F_1Q + (QR + RF_2) =$$
$$= (F_1Q + QR) + RF_2 = F_1R + RF_2;$$

we have the required conclusion, that being $F_1P + PF_2 < F_1R + RF_2$ and so the contact point P of a tangent to the ellipse is the solution of the Heron Problem for the tangent line and the foci and in this way the reflection property of the ellipse is established. It is now a simple exercise in opposite angles to see that a light ray coming from *outside of the ellipse aimed at one focus would reflect back from the ellipse on a line from the other focus.*

This reveals only the beginning of the story for it transpires that all three types of curves that arise from conic sections, those being ellipses, parabolas, and hyperbolas, each have their own characteristic reflection property with its own applications. The major application of the elliptical reflection property is certainly more devastating than poor Heron could ever have guessed as it is used critically to detonate thermonuclear explosions (H-bombs). By spinning an ellipse about its longer axis we generate a three-dimensional shape, similar in appearance to a rugby ball, that also has the reflection property, as the cross-sections of this *ellipsoid*, as it is called, through the principal axis are all identical ellipses that share the same pair of foci. The trick to setting off the fusion explosion involves a detonator at one focus of the ellipsoid, the energy from which is reflected back to the second focus in order to trigger the main device.

Cones, Curves, and Lamps

One game that mathematicians often play when searching for new ideas is to take an old idea, modify it in some minimal but significant fashion, and follow where it leads. If we replace addition by subtraction in the definition of the ellipse then we create a new class of curves known as *hyperbolas*. To be more explicit, we begin with a pair of fixed points, F_1, F_2 as before but this time we look to find the curve which consists of all points P the *difference* $PF_1 - PF_2$ of which is some fixed constant, a. (Remember that an ellipse arose when we took the corresponding *sum* to be a fixed number.) If $a = 0$ we are simply left with the line of points forming the perpendicular bisector of the line joining our two foci but otherwise we are rewarded with a genuinely new curve, the hyperbola. If we look at the opposite difference $PF_2 - PF_1$ and consider the curve consisting of all points for which this difference is a then the curve that arises is merely the reflection of the first curve in the perpendicular bisector of the line joining the foci. These two curves are actually regarded as separate branches of the same hyperbola, so that a hyperbola consists of two identically shaped curves (see Fig. 5.5).

A ray coming from one focus of the hyperbola strikes the surface and reflects as if it had come from the second focus. From this it is again an exercise in opposite angles to see that a ray coming from *between* the branches of the hyperbola and fired at one focus will reflect back *towards* the other focus. The ellipse and the hyperbola often exhibit behaviours that are inverse and complementary to one another, the reflection property being a case in point, and you will see other examples in what follows. The hyperbolic reflection property can again be interpreted as a Heron Problem involving the tangent line L at a point P on the curve and the two foci. However, to formulate this as a Heron Problem we require the two fixed points

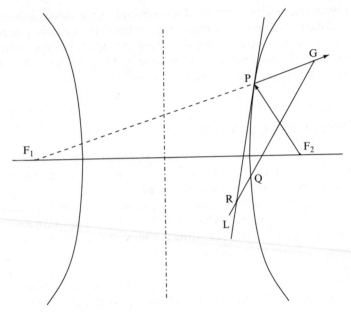

Fig. 5.5: Reflection property of a hyperbola

to lie on one side of the line L, which is evidently not the case here. To circumvent this difficulty we re-examine the Heron Problem (Fig. 5.2) and observe that the same solution applies if either of the points F_1, F_2 is replaced by *any* point along the ray from the optimum point P to F_1, or F_2 as the case may be. Bearing this in mind, extend the line F_1P to the *right* and choose any point G on that extension. The reflection property of the hyperbola then amounts to saying that P solves the Heron Problem for the tangent L and the points F_2 and G and in order to show this, after the fashion of the elliptic case, we would verify that $GP + PF_2 < GR + RF_2$ for any other point R on L, the details of which are recorded in the final chapter.

For a given pair of foci the collection of all ellipses with those foci forms a series of concentric ellipses covering the whole plane. The same is true of the family of all the hyperbolas with this foci pair in that they too cover the plane in this fashion: every point P in the plane lies on exactly one hyperbola of the family, that hyperbola being determined by the constant a, where $a = PF_1 - PF_2$. Moreover these families form what is known as an *orthogonal net of curves*, meaning that each of the ellipses meets each of the hyperbolas at right angles—in other words, the tangent of an ellipse is at right angles to the tangent to the hyperbola at the same point (see Fig. 5.6). This orthogonality effect is itself an artefact of the reflection properties of these curves. To see this recall that a ray from F_2 will reflect from the ellipse at the point P to the first focus, F_1, but that the same ray reflecting from the hyperbola will travel in the exact opposite direction away from F_1 through the line to P. This can only happen because the tangent to the hyperbola at P is at right angles to that of the ellipse at the same point. We say that the tangents are mutually *normal* or *orthogonal* to one another.

As has been mentioned briefly before, these types of curves arise through cutting a cone with a plane and indeed that was the way the curves were originally introduced by Menaechmus in Athens in

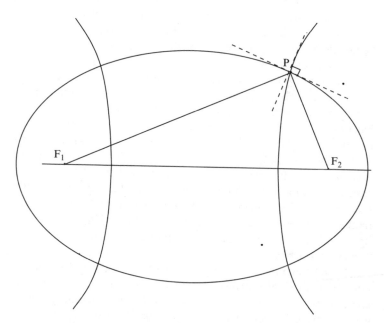

Fig. 5.6: Orthogonal ellipse-hyperbola pair

the 4th century BC. He was interested in these curves as a means to performing constructions which seemed to be impossible using Euclidean tools alone.

If we take a cross-section of a cone parallel to its base then the curve at the top of this truncated cone is, of course, circular. If we slant the plane of the cut, the curve that emerges is an ellipse. If we continue to slant the cutting plane the elliptical cross-section becomes increasing elongated until we reach the point when the cutting plane is exactly parallel to the slanting edge of the cone. The curve that then arises does not close up and can be shown to be a parabola, the curve of a projectile under gravity. The parabolic case is but a fleeting thing because as the tilt of the cutting plane continues to increase and goes past the parallel direction the resulting curve immediately changes character to that of a hyperbola. This phenomenon you will have seen before as the curve that separates light and shadow when a shaded lamp stands next to a wall. The light emerges from a point source in the bulb through the circular top of the shade, so producing a cone of light. The plane of the wall meets this light cone in one branch of a hyperbola which is made visible as the surface of the wall within the hyperbolic curve is lit while the rest of the wall is in the shadow of the shade.

To capture the second branch of the hyperbola we need to replace the cone by a *double cone* consisting of two identical right cones, known as *nappes*, with common axis and vertex, so that one balances precariously on top of the other. The introduction of this double cone approach was due to Apollonius of Perga in the 3rd century BC and is found in his extensive eight-book work *Conic Sections*. This definitive investigation superseded the relatively minor earlier works by Menaechmus, Aristaeus, and Euclid. Although the final book in the sequence is lost, this monumental work of pure mathematics came into its own nearly 2000 years later when in the hands of Isaac Newton it provided the framework for the explanation of planetary orbits.

Newton established through his mathematical physics that the underlying orbit of a planet or comet around the sun is that of a conic section. (Kepler had already shown that planetary orbits were elliptical.) Which particular type of orbit applied depended on the net energy of the orbiting object, which is the sum of two

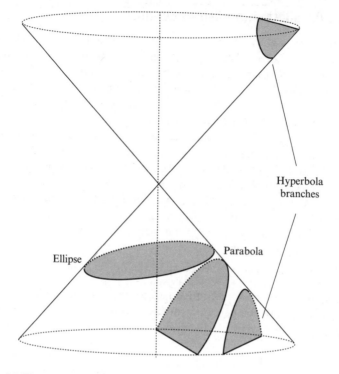

Fig. 5.7: Elliptical, parabolic, and hyperbolic conic sections

components: the object's *kinetic energy*, which is the energy it possesses due to its motion, and its *potential energy*, which is always a negative quantity due to its position in the region of the sun's gravitational field. If the overall energy of the object is negative then it will remain captured within the solar system, orbiting its ruler, the sun, in an elliptical orbit, but if its kinetic energy exceeds the magnitude of its potential energy its orbit will be one branch of a hyperbola—after passing by the sun it will escape the solar system never to return (as do some comets). The case of the parabola corresponds to the kinetic and potential energies in perfect balance, in which instance the object has a parabolic orbit. This transitional case does not occur in practice as it represents only a theoretical demarcation between the two basic orbit types of the ellipse and the hyperbola.

The Parabola: the Case of Equality

As mentioned above, Menaechmus introduced our three curve types as sections of a cone. They also arise in other distinct fashions. They may be introduced algebraically, but of that more later. Another construction shows that they arise purely out of considerations in plane geometry. This is the so-called *focus-directrix* approach which first appears in the *The Collection* of Pappus although it seems it was known earlier and Apollonius was also aware of focal properties of conics many centuries before. This method is most easily understood as applied to the parabola so we take that as our starting point.

We begin with a fixed point F, the *focus*, and a fixed line D, known as the *directrix*. Imagine a point P that starts midway between F and D and moves off in such a way that the particle always maintains an equal distance to the focus and the directrix. In symbols we are asking that,

for any point P on the parabola, $PF = PN$,

where N is the nearest point to P on the line D (see Fig. 5.8).

The path of the particle, which can equally move up or down the page, consists of a curve and that curve is a parabola as we shall

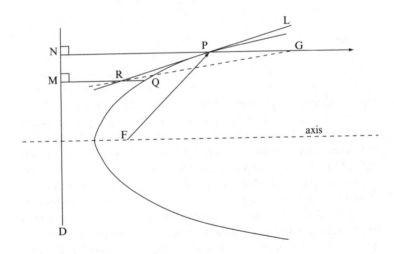

Fig. 5.8: Parabolic reflection property

define it here. The reflection property of the parabola is that rays emanating from the focus F all reflect parallel to the axis of symmetry of the curve. Conversely, rays coming in from the right and parallel to the axis will be reflected to the focus. We can interpret this in a similar fashion to what we have seen with the ellipse and the hyperbola by saying that the parabola has a second focus 'at infinity' in the direction of its axis and, with this way of looking at things, the parabola behaves as does the ellipse in that rays from one focus are reflected to the other. It is once again an exercise in opposite angles to check that a ray parallel to the axis coming from the *left* of the parabola reflects as if it had come from the focus or, equivalently, a ray aimed at the focus coming from the left of the curve is reflected parallel to the axis.

Rotating the parabola about its axis generates a surface of revolution known as a *paraboloid* which shares this reflection property and so has many uses: for example, the parabolic reflector in a car's headlights generates a strong parallel beam from the bulb placed at the focus, while the principle is used in reverse by the parabolic mirror of a reflecting telescope (an invention of Sir Isaac Newton) to focus parallel rays of light from a distant source such as a planet or galaxy outside of the Milky Way.

To prove the reflection property by reformulating it in terms of a Heron Problem we let G be any point to the right of the parabola and imagine a light ray travelling through G and parallel to the axis striking the curve at the point P. Let N be the foot of the perpendicular from D through P and G. Once again we assume what is physically true: the ray will reflect with equal angles of incidence and reflection to the line L tangent to the curve at P. We wish to demonstrate that the reflected ray passes through F, which is the same as saying that P is the solution to the Heron Problem for the tangent L and the points F and G. It remains therefore to test that $GP + PF < GR + RF$ for any other point R on L.

We first show that $GP + PF < GQ + QF$ for any other point Q at all on the parabola. Let M be the foot of the perpendicular from Q to the directrix. Making use of the defining property of the parabola ($PF = PN, QF = QM$) we discover the following inequalities:

$$GP + PF = GP + PN = GN < GM < GQ + QM = GQ + QF.$$

Now, taking Q to be the point where the parabola meets the line GR and using what we have just found allows completion of the argument as in the case of the ellipse:

$$GP + PF < GQ + QF < GQ + (QR + RF) =$$
$$(GQ + QR) + RF = GR + RF;$$

thus showing that P is the solution of the Heron Problem for the tangent L and the points F and G, as required to ensure the reflection property for the parabola. That is to say, rays parallel to the axis are focused to F or, what amounts to the same thing, rays coming from the focus reflect back parallel to the axis.

As mentioned at the beginning of this section, all conic sections types are realizable through the focus-directrix definition and this is done by the introduction of a positive constant called e, the *eccentricity* of the curve. With the focus F and directrix line D as above we consider the curve consisting of all points satisfying the equation $PF = ePN$, where $e > 0$ is fixed. If $e = 1$ we are of course back in the case of the parabola but if $e < 1$ the curve is no longer parabolic but closes up to form an ellipse, while if $e > 1$ the curve that arises is one branch of a hyperbola indicated in Figure 5.9.

Once the directrix and focus are chosen the curve is entirely determined by the value of the eccentricity. For $e = 1$ we have a parabola which instantly becomes an ellipse as e drops below 1. For values of e just below 1, that is, for high elliptical eccentricity, the ellipse is very large, but as e decreases the ellipse becomes more circular and the second focus of the ellipse moves continuously towards the fixed focus F and the ellipse becomes hardly distinguishable from a small circle centred at F. If on the other hand e increases past 1 the parabola immediately becomes one branch of a hyperbola and, as e increases further, the curve opens out and backs towards the directrix until for large values of e the curve would hardly be distinguished from the directrix itself.

The final approach to conics is through the equations that represent them. In order to proceed this way we need a coordinate system in which to work and this development is due primarily to René Descartes, the 17th-century mathematician and philosopher, and indeed it is from his name that we derive the word *Cartesian* for

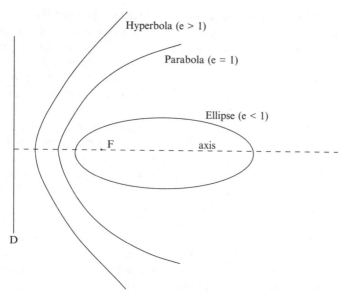

Fig. 5.9: Conics with shared focus

the familiar xy coordinate system. For example, a circle centred at the origin O of radius a is the set of all points $P(x,y)$ whose coordinates are related by the equation $x^2 + y^2 = a^2$. This in turn is just a statement of Pythagoras' Theorem for the triangle with hypotenuse $a = OP$ and legs parallel to the x and y axes. A straight line on the other hand can be represented by an equation of the form $y = mx + c$. The constant m is called the *gradient* or *slope* of the line and is a measure of its steepness while the number c is the value of the y-coordinate where the line meets the y-axis. (The exception is that the equation of a vertical line is specifed by the equation $x = a$, where a is the common value of the x-coordinate of each point on the line.)

The idea of coordinates to identify a position was certainly not novel in the 17th century, as it had been known to the Greeks and indeed the Romans and the Egyptians before that. Also, Apollonius derived most of his results on conics using the geometrical counterparts of Cartesian equations. It seems however that the algebraic method of studying curves could not develop until the advent of proper algebraic notation and the will to make uninhibited use of it, which was certainly a European innovation that took time to mature.

A potted history of this rise of algebra runs as follows. The Italian Renaissance of the late 15th and early 16th century set the scene. Pen and paper arithmetic as we are taught it came into its own: for example, the first known example of a long division sum is by Calandri in 1491.[2] In 1557 we see the first use of the sign '=' to denote equality in a book by the Englishman Robert Recorde, although the symbol he used has the two lines very elongated because, as he explained, the motivation comes from mimicking a pair of parallel lines which can hardly be more equal. This piece of notation soon caught on and the lines were quickly abbreviated in length to give the modern symbol for equality.

The first clear advance in mathematics since ancient times came in the first half of the 16th century when Scipione del Ferro, and later Tartaglia, Cardano, and his student Ferrari, showed how to solve third and fourth degree polynomial equations in the fashion of the familiar quadratic formula for finding the roots of second degree polynomials, which has been known, in effect, since the days of the ancient Sumerian scribes, perhaps close to 4000 years ago. These impressive achievements gave urgency to the development of new algebraic methods, although they were yet hampered by confusion regarding fundamentals, such as the status of so-called *imaginary numbers* (square roots of negatives) and to some extent the reality of negative integers was still doubted.

The credit for the establishment of algebra as we recognize it today unquestionably goes to the brilliant French mathematician François Viète who, in the latter part of the 16th century, emancipated algebra from the geometric style of proof by introducing the plus and minus signs and, crucially, introducing letters to stand for

---

2. No ancient European society developed a complete positional numbering system where the value of a numeral depends on its position within the number and full use is made of a zero symbol. By Archimedes' day the Greeks did have a positional system to the extent that Archimedes could demonstrate in *The Sand Reckoner* that he could write down a number greater than the number of grains of sand required to fill the universe. Early in the Christian era a complete positional system came into being in India with a symbol for 0 called *sunya*, the Hindu word for empty. It passed to Europe via the Arabs so our numeration system is called Hindu-Arabic.

arbitrary numbers that were then manipulated according to the laws governing arithmetic.[3] In this way algebra became a branch of mathematics that you could truly do something with and this is testified to by the observation that modern school students can use their algebraic facility to routinely derive results that would have staggered all the great thinkers of antiquity. Standard methods that might appear straightforward to us they would have regarded as a revelation of unbelievable power.

Mathematicians themselves are somewhat ambivalent towards using algebra to do geometry as it aims to replace thinking instead of stimulating it. The aesthetic bias that all mathematicians seem to have to varying degrees is that the use of coordinate systems is only for *calculation*. In order to *understand* the nature of geometry, a coordinate-free approach should be adopted as much as possible, so that our intuition is not clouded by the somewhat arbitrary coordinate systems imposed. At its heart mathematics is not a technical subject but is about ideas. For that reason its practitioners are resentful and suspicious of anything that introduces an arbitrary character into their investigations.

The algebraic method applied to conics is not however without elegance. Conic sections correspond to curves of arbitrary second degree equations in two variables, x and y. That is to say, ones of the form

$$a_1 x^2 + a_2 y^2 + a_3 xy + a_4 x + a_5 y + a_6 = 0$$

where the coefficients $a_0, a_1, \cdots, a_6$ are fixed numbers. Given such an equation, we can find another coordinate system in which the equation has a simpler form by rotating the axes through a suitable angle (which is a trigonometric function of the coefficients of the above equation) and then shifting the origin. In this way the equation can be expressed in a simple standard form

$$\frac{x^2}{a^2} \pm \frac{y^2}{b^2} = 1,$$

and we have an ellipse or a hyperbola according to whether the sign in the middle is plus or minus. The parabolic case is always peculiar

---

3. The $+$ and $-$ signs were of Germanic origin and Viète's algebra still fell some way short of modern notation: e.g. he would write A^3 as A cubus.

and generally simpler in that the equation adopts the form $y^2 = ax$ (or $x^2 = ay$). The constants a and b bear simple relationships to the conic in question: for the case of the ellipse the values of a and b respectively equal half of the horizontal and vertical axes of the ellipse. In the hyperbolic case the lines through the origin with slopes $\pm\frac{b}{a}$ are the so-called *asymptotes* of the curve: as we move along either branch of the curve away from the origin the curve itself straightens up and becomes ever closer to the asymptotes which therefore act as good linear approximations to the hyperbola in regions that lie far from the origin and the foci. For this reason, a deep space probe for example that is escaping the gravity of the solar system eventually flies away along a path that very much resembles and becomes ever nearer a straight line heading directly away from the sun (although the sun lies at a focus rather than the centre of the conic).

The eccentricity, e, of the conic may also be written in terms of a and b for, taking a to be larger than b, it turns out that:

$$e^2 = \frac{a^2 \pm b^2}{a^2},$$

where we take the positive sign for a hyperbola while the negative sign applies to the ellipse.

The correspondence of conics and second degree equations began the subject of what is known today as *Analytical Geometry* that dominates modern undergraduate mathematics to the extent that it and the associated methods of calculus represent mathematics in the mind of many a modern student. These findings on conics were published by Descartes in his *La Geometrie* in 1637, although the same series of discoveries was made independently by Fermat in the late 1620s. It is a straightforward matter, for example, to express conics given in focus-directrix form through equations governing their cartesian coordinates in a suitably positioned system of axes and their various geometric properties can then be unpacked and presented through analytical geometry.

6 ○ Covering the World

In 1852, Francis Guthrie, a mathematics student of University College London, asked himself how many colours are required so that a map may be coloured to ensure that neighbouring regions are coloured differently. He became convinced that four colours were enough for any map but could not prove it and so the problem was brought to the attention of his teacher, the famous logician, De Morgan. The problem looks so natural and simple it is difficult at first sight to believe that it could be very hard. A professional mathematician would be wary of admitting of having tried unsuccessfully to solve such a problem in case it turned out to be as simple as it looks. It was, however, a completely fresh problem and all the centuries of mathematics that had gone before were hardly any use at all in the face of this little question and so all investigators of the four-colour problem began on an equal footing, knowing practically nothing about how to tackle it.

Around this time it was beginning to be appreciated that questions about regions in the plane or volumes in three-dimensional space could be inordinately difficult to deal with in full generality, so it may not have taken De Morgan long to become convinced that this was a hard problem after all. He wrote to his friend, the leading Irish mathematician of the day, Sir William Rowan Hamilton,[1] in October 1852 and mentioned Guthrie's question, noting that the conjecture could be shown to be false if we could draw *five* regions in the plane that all bordered one another, although, he added, that did not seem possible.

That mathematicians were slow to recognize Guthrie's Problem as truly worthy of attention is attested to by the fact that the first article published on it in 1860 was anonymous (it is thought to be due to De Morgan), while Cayley's 1879 article 'On the Colouring of

---

1. Hamilton could read Greek, Hebrew, and Latin when aged 5 and by 10 was acquainted with half a dozen oriental languages. Friendships with Wordsworth and Coleridge led him to try his hand at poetry but he became one of the outstanding mathematicians of the latter part of the 19th century.

Maps', appeared not in a mathematics journal but in the *Proceedings of the Royal Geographical Society*. A. P. Kempe published a proof in the same year, on the strength of which he was admitted to the ranks of the Royal Society, but that was not to be the end of the story. In 1890 P. J. Heawood showed that Kempe's proof contained a fundamental error that could not be overcome, although the method of Kempe could be used to prove that no more than *five* colours would ever be needed to colour a map. Despite some progress, the problem stood defiant for another 86 years until Kenneth Appel and Wolfgang Haken announced in 1976 that the conjecture was true. Their proof was the first important example of a computer assisted proof in that they reduced the problem to over 1000 different cases that could only be verified through massive calculations performed by computer.

Many mathematicians did not like the look of this, declaring that it merely showed that the four-colour problem was 'not a good problem after all', meaning that, instead of being a source of new mathematical ideas, it turned out to be merely an enormous calculation that failed to shed much light on anything else. What is more, since the proof was so difficult to check, it was hard to judge how certain we could yet be of the answer. Indeed, should the proof turn out to be wrong, it could do more harm than good. Twenty-five years have passed however and no flaw has been discovered. Attempts have been made at refinement: in 1996, Robinson, Sanders, Seymour and Thomas tried to verify the Appel and Haken proof but 'soon gave up'. They embarked instead on their own proof and claim to have reached the same conclusion with fewer than half the number of cases and a much quicker checking procedure for each one. The basic approach is similar to the original and the proof is still computer assisted. It now looks like Guthrie, who never published anything on his own problem, was right from the beginning.[2]

Returning to De Morgan's first comment on the matter, it is not too hard to show that his observation regarding five regions on a sheet of paper is correct.

---

2. Guthrie became a mathematics professor in South Africa where he is also known for his contribution to botany, several rare species of flower bearing his name.

Can five regions be drawn in the plane so that each of them borders every other?

This really is impossible, although we first need to clarify what we mean by the word bordering. For two regions to share a common border we mean more than just a point—there needs to be a common boundary curve. If this were not stipulated then we would need more than four colours. To see this we may take a map of the USA and for our regions consider the four states of Arizona, Colarado, Nevada, and New Mexico (that do all meet at a single point, naturally called *Four Corners*), and the rest of North America as the fifth region. If one point can count as a common border, then all five regions border one another. Indeed it is not hard to see that Guthrie's Problem becomes easy and uninteresting under this interpretation, as if we take a pie and slice it into n pieces from the centre in the usual way, then all n regions of pie share a common central point so that if the centre were regarded as a common boundary then we would need as many colours as we have slices to colour this arrangement. Another proviso we need to make in order to have a real problem is that a region is a *connected* area of the globe consisting of a single piece—one country is not permitted to have an enclave totally enclosed by another nor is a country allowed to consist of two or more disconnected pieces as did Pakistan after the independence of the subcontinent and before Bangladesh became a nation in its own right. Once again, if we were to neglect this condition, then there is no limit to the number of colours that might be required for a map.

Returning to our question of the five regions, suppose now we have a map. Place an asterisk in each region and join two of these 'stars' by an arc if their regions have a common border as in Figure 6.1. For example, the star of the 'outside' region 1 is connected to those of regions 2, 3, 4, and 6 but not to 5 as region 5 does not have a border in common with region 1. The drawing of all these connecting arcs can be carried out without any two arcs (dotted lines) ever crossing one another—that is, two arcs will not meet except perhaps at one of the stars; this is because each arc may be drawn to lie in a four-sided figure the corners of which are the two stars connected by the arc and the two endpoints of the common border associated

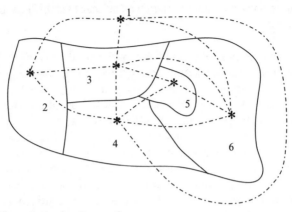

Fig. 6.1: Network of regional connections

with the regions of the arc—no other arc need enter this four-sided region. We call this collection of stars and arcs the *plane network* of the map.

Suppose now that we did have five regions on a map with each pair of regions sharing a common border. The plane network of these five regions would then consist of five stars and ten arcs (each of the *five* regions is joined to *four* others and each arc joins *two* regions giving $(5 \times 4)/2 = 10$ arcs), with an arc joining each pair in such a way that the arcs did not cross one another. However, this is an impossible object and it is not hard to explain why it cannot exist. If we name our stars *a,b,c,d* and *e*, then the arcs of the network described by the cycle $a \to b \to c \to d \to e \to a$ form a closed figure splitting the plane into an inside and an outside. This accounts for five and so there are still five more arcs to be drawn and so at least three of them are on the outside (with the rest inside) or at least three are on the inside of this figure. However, it is not possible to draw even two arcs inside the figure of five stars without them crossing, unless both begin at the same star, and then it is certainly not possible to draw a third inside without one unwanted crossing. There is no difference between the inside and outside of the figure as regards the validity of this argument—only two arcs can be placed outside the figure without crossings and these must have a common endpoint. It follows that, however we go about it,

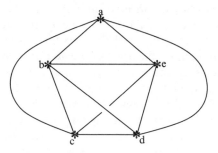

Fig. 6.2: Impossibility of five bordering regions

we shall not be able to draw that final tenth arc without crossing another arc already drawn. The best we could manage would be an effort resembling Figure 6.2.

And so we see that it is not possible to have five regions on a map any two of which share a common border. You should not imagine that this proves that the four-colour conjecture is true, for it only shows that, with any map, you will be able to colour any five regions without a colour clash but it is still conceivable that, if there were a lot of regions to colour, interactions between these sets of regions might somehow make it impossible to use the same four colours throughout.

Patterned Coverings

As soon as anyone thinks of covering a region, as you do when tiling a floor for example, they generally picture some kind of patterned arrangement, pleasing to the eye, as opposed to an arbitrary covering by blobs, which is what we see on a political map. The job can certainly be done with squares and rectangles in such a way that there need be no gaps left unfilled. Ancient mosaics reveal other possible tilings by equilateral triangles and regular hexagons, the latter type being discovered by honey bees in the course of their evolution to work for them in their hives. Covering by regular octagons does not work but leaves square gaps, a failure that leads to the idea of using *pairs* or *trios* of shapes to make your paving. Indeed there are eleven so-called *Archimedean tilings* of the plane, by which we mean tilings by convex polygons (plane figures, such as diamond shapes, in which all internal angles are less than 180°), not necessarily all alike (for example, the octagon-square tiling), but with identical arrangements at each corner, or *vertex*, as these points are known. Perhaps the prettiest of these is that involving a hexagon, two squares, and an equilateral triangle at each vertex, revealing a pleasing underlying circular pattern (see Fig. 6.3): at each vertex there meet one hexagon, two squares, and one triangle.

There are still other simple tilings based upon a single tile to be mentioned. It is easy to tile with any *parallelogram*, a four-sided figure where, just as in the special case of the rectangle, opposite sides are parallel but adjacent sides do not have to meet perpendicularly. Splitting each parallelogram along one of its diagonals gives you a covering of the floor with triangles and since two copies of *any* triangle can be joined to form a parallelogram, it follows that we can tile the plane with identical copies of any triangle. This suggests that the same may be possible using any four-sided figure.

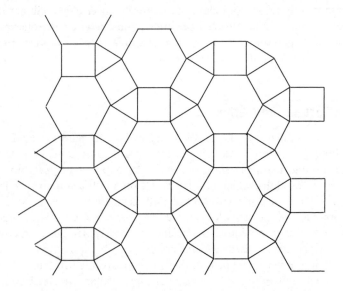

Fig. 6.3: An Archimedean tiling by hexagons, squares, and equilateral triangles

Can you tile the plane with copies of a single quadrilateral?

It is a fact that is not so generally known that the answer is yes—the tile does not need to have any special shape and does not even need to be convex. To make the point let us take the concave four-sided shape pictured in Figure 6.4.

We begin with a copy of the tile and form new tiles by taking the midpoint of each edge and rotating the tile through 180°. The tiles then fit together tightly without gaps or overlaps because every vertex is surrounded by each of the four angles of the quadrilateral which sum to 360°, that of a full turning, no matter what shape the figure has. (Working through an example with a paper cut-out is child's play—you'll find that your hands soon know what to do!) The success of this approach is also related to the well-known fact that the midpoints of any quadrilateral form a parallelogram and these underlying midpoint parallelograms themselves form a related tiling of the plane.

Since any four-sided figure will do to cover the plane without leaving gaps, what about five-sided figures? The regular pentagon with its five equal sides and angles is the simplest but it will not do. All our previous examples have the nice Archimedean property of having identical arrangements at each corner but the internal angle

Fig. 6.4: Tiling with a quadrilateral

of a regular pentagon is $540°/5 = 108°$,[3] which does not divide $360°$ a whole number of times—if we put three pentagons together at a vertex there is a gap, while four will not fit. Even if we abandon the Archimedean stipulation and allow messy tilings with vertices of some pentagons being placed along edges of others there is still no way of doing it, for the angle between an edge of one pentagon and the exterior angle of another is $(180 − 108)° = 72°$, leaving a gap that cannot be plugged by another pentagon. Hexagons do work of course, as three fit together nicely at a corner, but regular polygons with more than six sides cannot tile the plane either as their internal angles are all too large to fit three together at a vertex (or two together at a side) and so gaps are inevitable.

You may have read in the press in recent years about the so-called quasicrystal controversy—the discovery of crystals with a previously thought forbidden five-fold or *icosahedral* symmetry. Why should this be a surprise?

A region in the plane (or in three dimensions which is the case of general interest to crystallographers) is *periodic* if it is like a tiling in that it consists of just one motif R, of smallest possible size, known as a *fundamental region*, repeated over and over in all directions, without overlaps, and each copy R' of R in this tiling can be obtained from R by moving R in a certain direction—that is to say, R can be placed on top of R' without the need to rotate or reflect either of these tiles, and this translation takes the entire tiling back onto itself so that things look just as they did before. We can see this in our Archimedean tiling of Figure 6.3. The most eye-catching motif is the twelve-sided figure consisting of a central hexagon flanked by squares on all six sides with triangles filling the gaps between the squares. It is true that the entire tiling consists of copies of this one dodecahedron but each of these figures overlaps with four other identical ones so that the dodecahedron is *not* a fundamental region of the tiling in the way that term was just introduced. To discover a fundamental region imagine that you were laying this tiling on a floor by tiling one diagonal row of dodecahedrons after another. Upon completing one dodecahedron you will have already laid two

---

3. A convex polygon with n sides may be partitioned into n triangles with a common vertex showing that the internal angles sum to $2n − 4$ right angles—we subtract 4 to allow for the full turning at the common vertex.

of the squares and three of the triangles of the next dodecahedron so the fundamental region is the shape you require to *complete* the next dodecahedron, that is, a shape made up of one hexagon, three squares, and two triangles. Since each square and triangle appear in *two* dodecahedra a fundamental region consists of part of a dodecahedron incorporating just three of the six squares and two of the four triangles of the full dodecahedron. This yields an underlying 'clam shell' tiling by a certain decagon (a ten-sided shape) as pictured in Figure 6.5.[4]

The quadrilateral tiling of Figure 6.4 is also periodic. The quadrilateral itself is not a fundamental region of the tiling, rather it is known as a *generating tile* as the full covering can be produced using this tile if we allow for other motions—in this case a rotation is needed. A fundamental region has to generate the tiling by translations alone and to achieve this we need to take two adjacent tiles

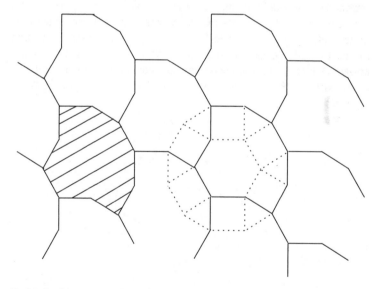

Fig. 6.5: Fundamental region of a tiling

4. Fundamental regions are not unique: for instance in Fig. 6.3 we may take as fundamental regions the parallelograms whose corners are the centres of adjacent hexagons: however, it can be shown that all fundamental regions of a periodic pattern have the same area.

with a common side—this produces a tiling by concave hexagons from which the original tiling can be recovered simply by partitioning each along a diagonal as shown in Figure 6.6.

The simplest tiling by squares exhibits four-fold symmetry in that when you rotate the pattern about a vertex it looks identical each time you turn it through 90° (and the same is true if we rotate around centres of the squares or the midpoints of their sides). The tiling by hexagons has six-fold symmetry as it looks identical after having been rotated through just 60° about the centre of each hexagon. Having six-fold symmetry it will automatically be three-fold symmetric also as rotation about the centre of a hexagon through 2 × 60° = 120° will also leave the pattern looking the same. The Archimedean tiling of Figure 6.3 also enjoys six-fold symmetry, but the underlying clam shell design, stripped of the extra decoration, shows no rotational symmetry. The square tiling, having four-fold symmetry, also has two-fold symmetry while a tiling by rectangles that are not square exhibits two-fold symmetry but no other rotational symmetry (except one-fold symmetry—any mosaic, turned through a full 360° about any point, returns you to your starting position). Five-fold symmetry then is the next one to look for.

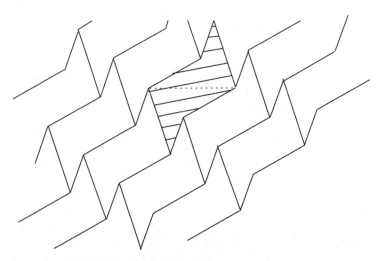

Fig. 6.6: Fundamental region of a quadrilateral tiling

Can a periodic tiling exhibit five-fold symmetry?

No, it is absolutely impossible for a periodic tiling to have five-fold symmetry or indeed n-fold symmetry about a point for any value of n greater than 6. To see this, suppose to the contrary that there were a plane periodic pattern that was the same, we sometimes use the word *invariant*, under a rotation about some point through the angle $(360/n)^\circ$ so that the pattern had n-fold symmetry. Since the pattern is periodic, there is an infinite number of such points of symmetry: let p and q be two of these centres that are separated by a distance, d, which is the minimum distance between any two centres of rotation.[5] Now since the pattern looks the same when rotated about p through an angle of $(360/n)^\circ$ (anti-clockwise or clockwise: in Figure 6.7(a) the first alternative is displayed), it follows that the centre of rotation, q, must be taken to be another centre of rotation, q', under this turning about p: this is only possible if n is no more than 6, for otherwise, since $360^\circ/n < 60^\circ$, we obtain two centres of rotation, q and q', that are closer together than d, which is impossible by our earlier choice of d as minimum (see Fig. 6.7(a)).

If on the other hand $n = 5$, then the angle $\angle qpq'$ in Figure 6.7(b) is $(360/5)^\circ = 72^\circ$, and q and q' are separated by a distance exceeding d,

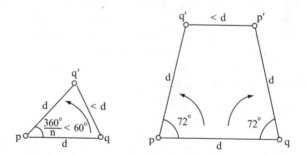

Fig. 6.7 (a) & (b): Restrictions on angles of rotation of periodic tilings

---

5. Readers sharp enough to ask the question 'how do we know there *is* a minimum distance d?' may like to consider this subtle query for themselves. This line of reasoning, known as Barlow's argument, does not rely on periodicity so much as the existence of this minimum separation d between centres of rotational symmetry.

so there is no immediate contradiction, but we soon spot another difficulty all the same. Since q is also a centre of rotation, rotating p about q through $72°$, this time in a clockwise direction, should take p to p', a new centre of five-fold symmetry in the picture, but then the two centres p' and q' are closer to each other than the minimum value d, again a contradiction.

The same argument is relevant in three dimensions for if we had a 'crystal', that is, a three-dimensional periodic filling of space that had five-fold rotational symmetry about some axis, we can apply the preceding argument to any plane at right angles to this axis of symmetry and arrive at the same contradiction, so five-fold symmetry of this type is just not possible, accounting for the fact that nobody had ever seen it, up until 1984 when crystals with icosahedral symmetry were created.[6] At first many scientists, including the double Nobel Prize Winner, Linus Pauling, thought that this must be some kind of misapprehension, as apparent five-fold symmetries had arisen before only to be explained away as not genuine. However, on this occasion it transpired otherwise. What had been discovered was a truly new and exciting phenomenon. It was not that the previous mathematical argument had been found wanting but that crystallographers have had to widen their view of the meaning of the word crystal from the traditional one of periodicity to the current working definition of that of a body that produces a crystal-like pattern under x-ray diffraction, that is to say, one with distinct bright spots. A new and unexpected frontier of physics and geometry has appeared quite out of the blue that itself has connections with non-periodic tilings, a topic to feature in the next chapter.[7]

---

6. The announcement came from Shechtman, Blech, Gratias, and Cahn in *Physical Review Letters* of 'a metallic phase with long-range orientational order and no translational symmetry'. The term *icosahedral symmetry* refers to regular icosahedra, three-dimensional solids consisting of 20 equilateral triangles, that have five-fold axes of rotation (and three and two-fold ones) and so, by our argument, cannot be packed to fill space periodically.
7. The story of quasicrystals is to be found in the excellent monograph *Quasicrystals and Geometry* by Majorie Senechal (Cambridge: Cambridge University Press, 1995).

The Arabic Legacy

The Golden Age of Greek mathematics begins about the time of Eudoxes who flourished in Athens around 350 BC and reached its peak in Archimedes who died in 212 BC. The Roman Empire and the Augustan Peace of the first two centuries of the Christian era were periods of relatively modest developments in the hard sciences. Although there are new ideas to be seen in the works of figures such as Apollonius, and later Heron, Diophantus, and Pappus, the momentum imparted by these and other individuals was not enough to sustain progress. In the end, it seems, the ancient Greeks ran out of ideas and the contribution from their culture ended.

Early medieval Europe rejected or at least neglected its pagan intellectual heritage to the extent that a great deal was lost. Fortunately, a more enlightened attitude sometimes prevailed in the newly emerging Muslim civilization. The caliphs of Baghdad were very keen to acquire all the worthwhile knowledge of both the Greek and Hindu worlds and many works from both sources were translated into Arabic in the 8th and 9th centuries. For example, the caliph al-Mansur (*c.*766) had the works of the leading Hindu mathematician of the 7th century, Brahmagupta, brought to Baghdad. This astronomical text, written in the year 628, also deals with aspects of mathematics. The caliph Aaron the Just who reigned from 786 until 808, known to us as Harun al-Rashid of the *Tales of Arabian Nights*, presided over the translation of Euclid's *Elements*.

It is often written that the role of the medieval Arabic world in the history of science in general and mathematics in particular was mainly one of preservation. The Muslim civilization eventually passed the mantle of ancient wisdom back to Europe when it awoke from its dogmatic medieval slumbers and was ready once more to advance the cause of human understanding.

To the extent that this is true it has to be seen in the context of the times. In the first millennium and more of the Christian era there is very little evidence of advances in mathematics anywhere in the world. In the Muslim nations the status of scientific learning was perhaps at its most secure and here we see a pattern somewhat similar to that under the Roman Empire where the works of antiquity were available to a privileged few and occasionally we see progress made on one front or another by an outstanding individual. For example we may cite the poet and mathematician Omar Khayyám who, at the turn of the 12th century, was a paid scholar in the patronage of a caliph. In addition to his poetry that has been translated into scores of languages[8] he carried out mathematical research into the solution of cubic and higher degree equations. Despite the fact that *algebra* is an Arabic word,[9] algebraic development was hampered by a total lack of notation so that Omar Khayyám sought geometric solutions as represented through intersections of a parabola and a hyperbola. This can all be formulated in modern notation in terms of equations and substitutions but Khayyám had no access to this language and he rejected the idea of negative numbers so he faced severe handicaps that he did well to overcome. He also considered the status of Euclid's Fifth Postulate and was led to consideration of the Saccheri Quadrilateral (Fig. 3.12) over six centuries before it again emerged in Europe. Unfortunately it seems there were not enough professional scholars to generate the momentum for sustained advances in the sciences.

If we look to India and China we find the history of mathematics there extremely obscure. They do however offer glimpses at their earliest stages that are similar in nature to that seen in the very ancient civilizations of Mesopotamia and elsewhere. For example, the Chinese classical work *Nine Chapters on the Mathematical Art* was probably written about 250 BC and consists of 246 specific problems drawn from all walks of life from surveying and engineering through to taxation. In Chinese mathematics there is a particular fondness for number patterns and here we see the first magic square: it seems that it truly was magic as we are told it was brought to a man by a turtle from the River Lo (Fig. 6.8): the numbers 1 through to 9 are cunningly arranged so that each line, whether it be horizontal, vertical, or diagonal, sums to 15.

Although magic squares are not very important, number arrays, or *matrices* as they are called, are a major ingredient of modern algebra and so it is particularly impressive to the modern

---

8. The moving finger writes, and having writ moves on...
9. The word derives from the book *Al-jabr wa'l muqabalah* written by al-Khowarizmi from whose name comes our word *algorithm* for a list of instructions for carrying out a computational task.

4	9	2
3	5	7
8	1	6

Fig. 6.8: The first Magic Square

mathematical eye to see an early anticipation of matrices in the ancient Chinese literature, for the author of the *Nine Chapters* solves a set of three linear equations in three unknowns by the modern method of column operations on the corresponding matrix array of coefficients. Linear algebra is a staple of modern mathematics and its calculations are based on manipulation and multiplication of such matrix arrays. It has become a most important branch of applied mathematics and its use is very widespread in both economics and physics although its evolution is recent.

The development of linear algebra is a feature of the late 19th and early 20th centuries and at first was considered an obscure branch of theory. A particularly uncomfortable feature of matrix algebra was the non-commutativity of matrix multiplication: given two matrices, *A* and *B*, the two matrix products, *AB* and *BA*, are generally different matrices. This 'unnatural' behaviour would have seemed offputting to non-specialists at first who would therefore prefer to ignore the topic entirely. When matrices arose irresistibly in quantum mechanics this caught some of its leading practictioners unawares. There is a letter from the theoretical physicist Max Born to Einstein where he is speaking of his work with Werner Heisenberg in 1925. He writes with bitterness of his colleague, 'in those days he actually had no idea what a matrix was. It was he who reaped all the rewards of our work together such as the Nobel Prize.'[10] Nowadays, however, matrix algebra is taught widely in schools and elsewhere and even the most mathematically narrow-minded have learnt to shrug off the strange nature of the multiplication in view of a host of practical uses for matrices. Again, we see that people can soon

---

10. Max Born was eventually awarded the Nobel Prize for Physics in 1954 for his work on the probabilistic interpretation of quantum mechanics, to the evident relief of all concerned, including Heisenberg.

become accustomed to a new mathematical development, even one that initially seems very strange, once it becomes familiar and especially if it proves useful. This all serves to make the achievement of the author of *The Nine Chapters* all the more striking.

Some later Chinese developments are also impressive. For example, the 13th-century mathematician Chu Shih-chieh of the later Sung Dynasty introduced an iterative method he called *fan fa* for finding successively better approximations of the solutions to quadratic and higher degree equations. In Europe this technique bears the name of Horner, who lived some 500 years later. Both Chu Shih-chieh and Omar Khayyám seem also to have been aware of the Binomial Theorem, the coefficients of which occur through what is now referred to as Pascal's triangle, named after the outstanding French mathematician and philosopher of the 17th century, Blaise Pascal. Indeed the late medieval period saw many tantalizing hints of future developments and a great variety of mathematical problems was tackled in an *ad hoc* and naive fashion, with little or no use of an abbreviating notation, to the extent that numbers themselves were sometimes written out in words. However, it could not be claimed that there was any systematic study of mathematics in its own right in China or elsewhere all throughout medieval times. In Chu Shih-chieh we have but another example of an exceptional individual who took what mathematics was available to him and embellished it with his own clever innovations. Neither he nor his subject had any real security for he lived as a wandering scholar and some of his own works were lost in China itself for centuries after.

None the less the period of Arabic hegemony did see some important developments, especially in trigonometry. The story of this subject is long and complex because knowing the relationship between side length and angles of triangles is of basic importance in measurement and so the topic was studied all throughout history, although in an erratic fashion, usually with particular applications firmly in view. The Greeks did not develop trigonometry as much as they could have as they too saw it more as a practical tool, a handmaiden of astronomy, undeserving of the status of pure geometry that merited a thorough investigation in its own right. For that reason it is not surprising that an important individual contribution to the subject came from the astronomer Hipparchus around 140 BC. He constructed a very accurate table of lengths of chords of a circle as a function of the angle subtended. This is equivalent to tabulating what we would call the sine function, although the origin of the word *sine* is Hindu and the earliest table of sines (half chords) comes from India in the 4th century. These were devised not through direct measurement only but by use of recursions for relating the sine of one angle to another. In the second century AD there were two major Greek contributions, those being the works of Menelaus and later Claudius Ptolemy, both of Alexandria, and there may have been some influence of Alexandrian scholarship on developments that followed in India.

Later Muslim mathematicians regarded themselves first and foremost as astronomers and so were very interested in both plane and spherical trigonometry. The Arabs took Greek trigonometry and applied Hindu forms to discover new functions and formulae. In particular the Muslim scholars made use of all six trigonometric functions and introduced the crucial *product to sum* formulae for the cosine and other functions that were introduced in Europe in the 16th century (before the advent of logarithms) to convert multiplications into additions in complex astronomical calculations.

Of particular interest to us here is the Arabian appreciation of the nature of tilings. This marriage of the practical arts and mathematics reached a new height in the decorative patterns of the Muslim world. Here we see a phenomenon that we also witness in the quilt designs of the American Amish society where constraint on what is acceptable decoration can act as a stimulus to creativity and the systematic development of an art form. Since depiction of the human face was forbidden in Muslim art there was a concentration on what was possible in geometric patterning and a thorough appreciation of symmetry was a by-product.

The Seventeen Symmetries of Alhambra

What makes a tiling or wallpaper pattern periodic is its translational symmetry. It is always possible to find two shifts of the pattern, in different directions, that take the pattern back on to itself and any such shift can be accomplished by combining these two basic translations. The two directions of translational symmetry are not necessarily at right angles to each other, as can be seen in our clam shell tiling of Figure 6.5 where the two translational directions can be taken to be directly up and down the page and on a line from bottom left to top right making an angle of 60° with the vertical. Any of the clam shells can be shifted on top of any other through use of these two basic motions and the pattern will appear identical to how it was before they took place. Inspection of our Archimedean tiling of Figure 6.3 soon reveals many more symmetries—the pattern has six-fold axes of rotation about the centre of each hexagon for example and the tiling may be reflected in the perpendicular bisector of the sides of each square. Some patterns have other forms of symmetry known as glide reflections that involve coupling a translation and a reflection in tandem to reveal a new symmetry of the pattern. It would seem there is no end to the complex symmetries possible in wallpaper and tiling patterns but this is not the case. It turns out that the feasible symmetries of a periodic tiling can be classifed into exactly seventeen types and no more. What is more, all seventeen possibilities feature in Arabic tiling designs, strongly suggesting that they appreciated this classification even if they did not have the mathematical language to write about it. These designs could be seen throughout the Arab world and in particular in the Alhambra Palace in southern Spain, the most beautiful monument of Arab civilization in Europe. This was the place where the Moorish king could live in marvellous surroundings most approaching the Koran's description of Paradise.

A hint that the symmetries of a pattern could be limited to a certain number of types arose before in the discussion of rotational symmetry. We say a pattern has *n-fold rotational symmetry* about a point if, when turned through an angle of $(360/n)°$ degrees about that point, it looks exactly the same—each motif, however far from the centre of rotation it may lie, is taken to rest upon another identical motif in exactly the same attitude so that the pattern appears unchanged. We saw that the only values of n possible were 1, 2, 3, 4, and 6. To understand the nature of the classification into the seventeen types it is first best to experience them through simple examples and those featured here are based on an article by Doris Schattschneider that appeared in the *American Mathematical Monthly* in 1978. Let us begin with the pattern in Figure 6.9.

The tiling has the inevitable translational symmetries of a periodic pattern but no others. Rotating the pattern about the centre of a rhombus (square parallelogram), or about a vertex, or the midpoint of a side, will move the underlying grid back onto itself but would turn each of the decorative shapes resembling the numeral 6, upside down. No motion involving a reflection can work either as that would result in all the curved figures being written backwards.

We shall introduce our remaining examples by increasing order of rotation. The pattern of Figure 6.9 has *order of rotation* of 1, meaning that the pattern has no rotational symmetry except for the trivial one of a complete turning about any point through 360°. The next pattern, Figure 6.10, also lacks rotational symmetry but does have obvious axes of *reflection* in any of the vertical lines of the tiling so it represents a second new symmetry class of periodic tilings.

Our third curious pattern, as shown in Figure 6.11, is somewhat different in nature. Once again we have a tiling with no rotational symmetry—rotation of the pattern around the midpoint of a side of a square would work except that the rotation places each figure 6 on top of another with the opposite orientation. Since the curved figures do occur in oppositely oriented pairs there would seem some hope of finding an axis of symmetry for a reflection but such an axis would have to lie along the side of a square, or pass perpendicularly

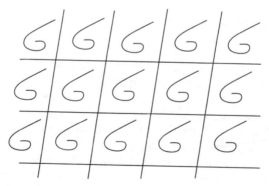

Fig. 6.9: Simple translational symmetry

Fig. 6.10: Reflectional but no rotational symmetry

Fig. 6.11: Pattern with glide reflection but no reflectional symmetry

through a side, or be a diagonal of the underlying pattern of squares and none of these alternatives do the trick because of the placements of the curved figures within each square.

The pattern does none the less possess a new symmetry of a type we have not seen before. We can map the pattern back on to itself by first moving everything along one square and then reflecting the resulting shifted pattern in the dotted line in the picture. This new type of symmetry, a translation followed by a reflection in a line parallel to the direction of the translation, is known as a *glide reflection*.

To be sure, the pattern in Figure 6.10 also possesses glide reflections as, for instance, if we shift the pattern one square down and follow this by a reflection in any of the vertical lines the pattern will again look the same. The difference between this and what we see in Figure 6.11 is that, in this latter picture, both the translation and the reflection that we used to create the symmetry were not themselves symmetries of the tiling: the tiling of Figure 6.11 does *not* look the same after it has been shifted one square to the right, and the same applies to reflection in the dotted line. We say that the pattern in Figure 6.11 has a *non-trivial* glide reflection, that is to say, one that does not arise simply through translations and reflections of the tiling—indeed we have already seen that this tiling has no axes of symmetry.

Having come this far it is time to take more careful stock of our position. Since we have discovered a new type of symmetry perhaps we should ask what kind of symmetries can arise for a periodic tiling? There are only four types possible: translations and rotations, the so-called *direct symmetries*, and reflections and glide reflections, which are classed as *indirect* symmetries for they reverse the sense of a figure, turning a right-handed glove for instance into a left-handed one.[11] Given two or more symmetries of our tiling we can apply one after another and the result will always be a symmetry of the pattern as each application of a symmetry leaves the figure looking the same as before. Mathematicians refer to the entire collection of symmetries of an object as its *group* of symmetries. Groups are not merely collections but have the special properties that we have seen of the families of symmetries of the preceding examples: two symmetries can be combined to form a new one, and the reverse of any symmetry is also a symmetry of the tiling—the reverse of a translation is the reverse shift in the opposite direction, the reverse of a clockwise rotation is an anti-clockwise rotation around the same point, while a reflection is the reverse of itself. A glide reflection has as its reverse (the word used in mathematics is *inverse*) another glide reflection that uses the same reflection but the opposite translation of the original symmetry.

---

11. This and the claims that follow are not hard to demonstrate: see Chapter 8.

These four fundamental operations of translation, rotation, reflection, and glide reflection are known as the *isometries* or the *rigid motions* of the plane: they are the only operations on plane figures that will always leave the distance between any two points unchanged (and, in consequence, leave the angles between lines invariant as well). The collection of all isometries form a group so that combining one isometry with another yields a new isometry. However, no matter how many times this is done, the net effect will be the same as carrying out just one of the four basic types.[12] For example, combining two reflections will yield a rotation about the point of intersection of the two reflection lines through an angle equal to twice that which lies between them, provided of course that the lines meet; if they are parallel the effect is one of a translation through twice the distance separating the lines in the direction perpendicular to them both. Combining two direct isometries will always yield another direct isometry as neither isometry can change a right-handed figure into a left-handed one. Similar reasoning allows us to say that the net effect of an indirect isometry with a direct one will yield an indirect isometry while two indirect isometries will result in a direct isometry. In this behaviour we see the pattern of adding even and odd numbers with the direct isometries corresponding to the evens while the indirect isometries correspond to odds. (Or, if you prefer, the pattern of signs associated with multiplying positive and negative numbers.)

Referring all this back to the three examples we have looked at so far we see that the group of symmetries of Figure 6.9 was as simple as possible, containing as it does only translations. To generate all possible symmetries for the tiling of Figure 6.10, however, we need to adjoin two reflections, while the symmetry group of Figure 6.11 requires glide reflections in order to produce all possible symmetries. There remains one other possibility for a tiling with no rotational symmetry, namely an example that possesses reflections and non-trivial glide reflections at the same time. This is realized in the design featured in Figure 6.12. The diagonals are reflection axes while lines of glide reflection, one of which is marked, lie parallel

---

12. In fact it can all be done with mirrors: any isometry of the plane can be achieved by three or fewer reflections, as shown in the final chapter.

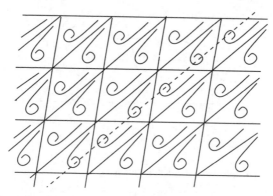

Fig. 6.12: No rotational symmetry but both kinds of indirect isometries

and halfway between adjacent lines of reflection: move one unit up and to the right and then reflect to see this in action.

This exhausts the possibilities for symmetry groups that do not admit rotations.[13] Next we look at patterns that admit rotations of order 2, that is to say, are symmetric with respect to a half turn about a certain point, but do not have rotational centres of higher order.

The following pattern, Figure 6.13, clearly has half-turn rotational symmetries about the points where the decorative curves cross the grid pattern and the points where the grid lines meet but does not admit rotations of orders 3, 4, or 6, nor does it allow for indirect symmetries because all the decorative curves share the same sense that would be reversed by any reflection. Our quadrilateral tiling of Figure 6.4 is a member of the same symmetry group.

Figure 6.14 also reveals a pattern with half-turn symmetry about the corners of each rectangle but in addition it offers reflections along all of the lines. This pattern does not have glide reflections other than those that can be generated by its translations and reflections.

---

13. The reader should realize that this statement has not been *proved* here, however plausible it might seem. The advantage of the group approach is that it classifies the families of symmetries within a framework that allows us a precise discussion that not only finds all symmetry types but makes it possible to show that there are no others, and so we are sure that the artists of Alhambra did not overlook any possibilities.

Fig. 6.13: Half-turn symmetry but no indirect isometry

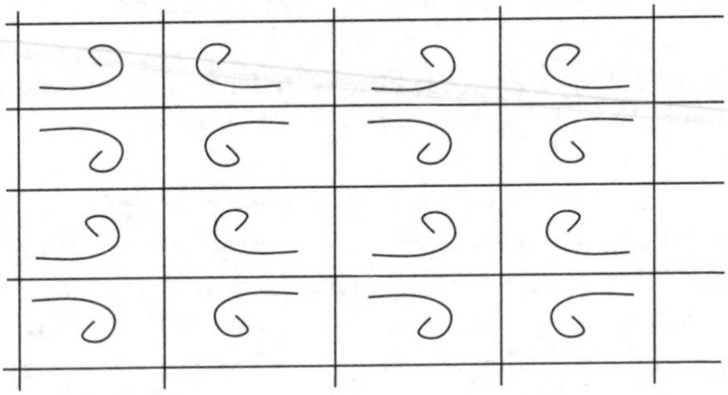

Fig. 6.14: Pattern with half-turn and reflection symmetry

The tiling of Figure 6.15 also has rotational symmetry of order 2 at the corner of each square but in contrast to our previous example has no reflections, although it does exhibit glide reflections. Once again one of the glide reflection axes is dotted and we see that the picture reverts to the same appearance should we shift the lattice framework one tile to the right and then reflect in the line indicated.

A pattern with rotational symmetry of order 2 can admit both types of indirect isometry, reflection, and non-trivial glide reflections, and our first example of this is shown in Figure 6.16. The picture involves the same features as the previous example and has half-turn symmetry about the midpoints of the vertical sides of squares (but not about the corners). However, the horizontal lines of

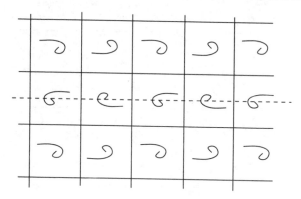

Fig. 6.15: Pattern with half-turn symmetry and glide reflection

Fig. 6.16: Pattern with half-turn symmetry, reflection, and glide reflection

the picture are now lines of reflection and a glide reflection results through translating the picture down one square and then reflecting in a vertical line. Since neither this translation nor reflection is a symmetry of the pattern, we have a genuine, that is to say, a non-trivial, glide reflection for the pattern, so the tiling represents a new class of symmetries.

Our next picture, Figure 6.17, is also one of a periodic tiling with half-turn symmetry as well as both reflections and non-trivial glide reflections. This time we see centres of half-turn rotation at every corner of the grid, including the intersections of the diagonals, along

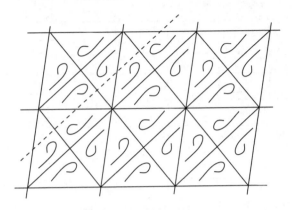

Fig. 6.17: Half-turn, reflection, and glide reflection symmetry (type II)

with the midpoints of the sides of each rhombus (the four-sided figures with sides of equal length which comprise the grid). The diagonals are lines of reflection and parallel to each line of reflection lies a line of glide reflection, one of which is identified by the dotted line in the picture. However, the symmetry groups of the tilings of Figures 6.16 and 6.17 are not the same: the reflection axes of Figure 6.16 are all parallel to one another while in Figure 6.17 we see mutually perpendicular reflection axes of the pattern, which is enough to distinguish between these two symmetry types despite the fact that they both have rotational order 2 and come complete with reflections and non-trivial glide reflections.

This exhausts the possibilites for the symmetry groups of patterns with half-turn symmetry so we next turn to patterns that have 3 as their highest order of rotational symmetry. That of Figure 6.18 has no indirect symmetries as all the 'sixes' are right-handed.

It turns out that any tiling with order of rotation 3, 4, or 6 that has indirect symmetries must possess both reflections and non-trivial glide reflections and, similarly to the final pair of examples of patterns with half-turn symmetry, there are two symmetry groups in this category in the case of symmetry of order 3, the first of which is represented by the pattern of Figure 6.19.

Every intersection point of the underlying hexagonal grid is a centre of three-fold rotation and every line on the grid is a reflection

Fig. 6.18: Simple rotational symmetry of order 3

Fig. 6.19: Rotational symmetry of order 3 with indirect isometries

axis. Halfway between an adjacent pair of parallel reflection lines we find a line of glide reflection, one of which is dotted on the diagram: to execute the glide reflection move the pattern to the right through one and a half rhombic units and then reflect. This pattern has all three-fold centres lying on some reflection axis, a feature that distinguishes its symmetries from that seen in Figure 6.20.

Fig. 6.20: Rotational symmetry of order 3 with indirect isometries (type II)

In Figure 6.20 we see an underlying hexagonal grid and where any two of the marked grid lines meet defines a three-fold centre of rotation. Again a line about which a non-trivial glide reflection may act is indicated. The difference between the symmetry groups of this pattern and the one before is manifested through the fact that not all the three-fold centres of rotation lie on lines of reflection: the horizontal line of centres through the middle of the diagram does define a reflection line for the pattern but the four other centres of three-fold symmetry in the body of the picture do not lie on axes of reflection. Midway and parallel to any adjacent pair of lines of reflection lies a line of glide reflection: if the picture is moved right so that each vertical line is mapped partly on to the next one, and then the reflection in the dotted line is carried out, the picture looks the same as before.

The previous three examples, Figures 6.18–20, represent all the possibilities for symmetry types where the highest order of rotation is 3. The behaviour as regards four-fold symmetry is similar: there is one type which exhibits no indirect isometries (Figure 6.21) and two distinct types that exhibit both reflections and glide reflections. In the first case (Figure 6.22) all the four-fold centres lie on reflection axes but in the second case not (Figure 6.23).

In Figure 6.22 the intersection of diagonal lines mark centres of four-fold symmetry and the picture may be reflected about every line. The grid intersections not involving the diagonal lines are centres of two-fold symmetry only and lines of glide reflection join these centres: again one is marked on the diagram and the glide reflection is executed by moving parallel to the line from one square

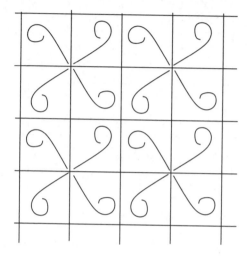

Fig. 6.21: Simple four-fold symmetry

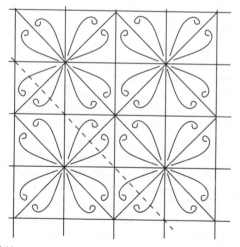

Fig. 6.22: Four-fold symmetry with indirect isometries

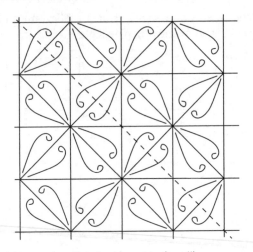

Fig. 6.23: Four-fold symmetry with indirect isometries (type II)

to the square one below and to the right and then reflecting in the dotted line.

In Figure 6.23 once again the diagonal lines are lines of reflection but only centres of half-turn symmetry are found on them—the four-fold centres lie at the other intersections of the grid lines and the axes of glide reflection are the diagonals joining four-fold centres of rotation, one of them being marked as a dotted line: again if we move in that direction one square down and to the right and then reflect in the dotted line we recover the pattern.

There remain but two symmetry types left: those patterns with six-fold centres of rotation. It is easy to spot the six-fold centres of symmetry in Figure 6.24 (interspersed with three-fold ones) but again any possibility of indirect isometry is stymied, this time by the presence of the J-like figures as they are all right-handed. On the other hand our final picture of the chapter, Figure 6.25, exhibits six-fold symmetry together with both kinds of indirect isometries. It represents the same group of symmetries as does our Figure 6.3, the Archimedean tiling. This is the richest of all the seventeen symmetry types of tilings of the plane. The reader should be able to identify six-fold centres of rotations as well as three-fold and two-fold

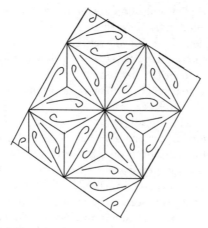

Fig. 6.24: Simple six-fold rotational symmetry

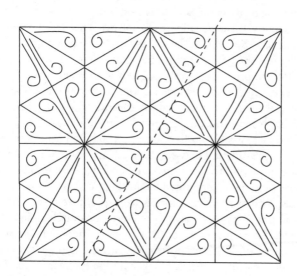

Fig. 6.25: Six-fold symmetry with indirect isometries

centres. Axes of glide reflection run between the two-fold centres and one is indicated in the diagram.

This completes our list of representatives of the seventeen groups of plane periodic patterns. Armed with this knowledge, you should be able to identify the symmetry type of any floral pattern or wallpaper design that you come across, or, should you visit Alhambra, that of an Arabian fresco.

The greatest 20th-century expositor of the art of plane tilings was the reknowned Dutch engraver M.C. Escher. Although claiming to have no mathematical training, his work shows an enviable appreciation of the nature of space not only from the Euclidean perspective but also of hyperbolic, elliptic, and projective geometries (the geometry of what is seen). His impossible constructs, such as stairways simultaneously leading both upwards and downwards, are the subject of many a student wall poster. Some of his work, such as his ants following each other endlessly around a Möbius strip (a strip joined up with a half twist so that it has but one edge and one side), could exist in reality while others have an ethereal, dreamlike quality. It is his appreciation of tilings that is extraordinarily acute to the extent that his work *Ghosts* can be interpreted as a solution to the eighteenth of David Hilbert's problems. This list of 23 problems was set in 1900 by Hilbert, the leading mathematician of the day, for researchers of the 20th century. The relationship between *Ghosts* and the eighteenth problem is explained in Majorie Senechal's *Quasicrystals*. Escher, free from the restrictions imposed on his medieval Muslim predecessors, managed to ingeniously work animal and human figures into his tilings. One favourite, *Knights on Horseback*, may seem at first sight to have only translational symmetry but by ignoring the distinction between the dark and light horsemen we see that the symmetry group admits parallel glide reflections and so is a member of the type represented by our Figure 6.11.

It seems however that Escher's work did have a genuine mathematical impetus. According to Doris Schattschneider, Escher 'struggled for several years to produce animated interlocking designs, with very primitive results'. The key moment for Escher seems to have came through a paper by the Hungarian mathematician George Polya (*Über die Analogie der Kristallsymmetrie in der Ebene*, Z. für Kristallographie, 60 (1924), 278–82). Schattschneider goes on to say, 'In examining Escher's notebooks, this author discovered that he copied in full the paper by G. Polya which outlines the important properties of each of the groups and includes a chart of illustrative designs.' It would appear that this was what allowed Escher's efforts to blossom and so we have an example of mathematics informing art not only in broad outline but in critical structural detail. To see other examples of such deep influence of geometry on art we would have to return to the Renaissance when a very precise analysis of perspective was undertaken. In that period we have examples such as Leonardo's work analysing the manner in which light falls on a sphere which some say manifests itself in the *Mona Lisa*.

Of course our world is not flat and so perhaps shapes that can fill space are ultimately more important than those that can tile the plane. As we pass from two- to three-dimensional space the number of so-called crystallographic groups increases from 17 to 230, all of which were discovered by 1891 by Fedorov and, independently, by Schoenflies. Mathematicians, as always, were the first to go into higher dimensions and by 1912 Bieberbach and Frobenius were able to prove that, for *any* dimension, the number of classes of crystallographic groups is always finite, that is to say, there is a limited number of them.

Behaviour in lower dimensions can often be understood as arising from projection from a higher dimension so it would be wrong to presume that knowledge of higher dimensional crystallographic groups would not be of interest in the real world. For example, so-called modulated crystal structures are best understood in this way, as kinds of three-dimensional shadows of a four-dimensional world. For that reason, a reference table listing the four-dimensional crystallographic groups, and there are 4895 of them in all, has been drawn up. The walls of Alhambra contain the seed of a big idea, for they point the way to the discovery of just what structures are possible in the universe we find ourselves in.

7 ○ Possible and Impossible Constructions

Constructions with a straight edge and compasses formed a cornerstone of mathematical training for many hundreds of years until the latter part of the 20th century. They are now only taught in passing and are at best a peripheral part of school geometry. This is a pity as they are really quite fun—many people who are not at home with other parts of mathematics, notably algebra, find to their surprise and delight a natural ability at this aspect of geometry. On the one hand, the topic itself is visual and tactile in a way that mathematics often is not, while on the other it embodies the geometric ideal as it explores what can be built and transferred from one figure to another without the intervention of coordinates and scales. There is no corner of mathematics that represents such a happy marriage of practical skill and pure mathematical ideals.

Despite the severe constraints imposed by the Euclidean tools, perfect models of complex geometrical objects can be approached using only compasses and straight edge, as we shall shortly see. Another pleasing aspect of the subject is that it lets students develop their skills through discovering new constructions themselves. Once some basic constructions have been demonstrated, such as bisecting a line, duplicating an angle, and constructing lines perpendicular and parallel to a given line through a given point, it is not difficult to devise new constructions. However, not all interesting constructions are easy to find, as we shall see, for example, in the case of the regular pentagon.

Showing that certain constructions are impossible is generally harder than obtaining positive results, as the former requires a precise algebraic description of the gamut of all possibilities before we can decide what lies within the realms of the possible. Such ground rules were not fully appreciated and formulated until the 19th century. For example, it is impossible to 'square the circle', that is, given a square, there is no Euclidean construction leading to a circle with the same area, a fact not proved until the 1880s. It is, on the other hand, an easy matter to 'square the rectangle' as we shall see shortly.

The ancient Greeks had a great respect for such constructions: they saw in them an aesthetic perfection manifested in this instance by the ability to transfer the precise area of a rectangle to that of a square without the intrusion of measurement and its intrinsic error. This held a great appeal to the classical mind that strove to separate the ideal from the mundane and distil its essence, an attitude that was the basis of Platonic philosophy where only the underlying ideal objects were regarded as truly real. This particular example is relevant to what follows so, taking some of the basic constructions for granted for the moment, we show how it can be done.

Given a rectangle $ABCD$ as in Figure 7.1 we extend the longer side to a point E so that $BE = BC$. Find the centre, O, of the line AE by bisection (see Fig. 7.4), draw the semicircle with diameter AE, and extend the line BC so that it meets the semicircle at X. The square with side BX then has the same area as the original rectangle. (To construct the square, extend BE and mark off, using your compasses, the length BX on this extended line segment to obtain the second side of the square; constructing suitable right angles then allows the complete square to be drawn.)

That the two areas are equal is a consequence of Pythagoras' Theorem. Let r denote the radius of the circle, let $BX = a$ and $BC = BE = b$. From the right-angled triangle $\triangle OXB$ we get

$$OX^2 = OB^2 + BX^2 \rightarrow r^2 = (r - b)^2 + a^2.$$

Taking all terms in r to the left and simplifying (either by expanding $(r - b)^2$ or using the difference of two squares) we obtain $a^2 = b(2r - b)$, which just says that the area of the constructed

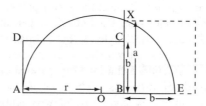

Fig. 7.1: Squaring the rectangle

square, a^2, equals the area of the original rectangle $ABCD$, (as $AB = 2r - b$).

When we erect a line of unit length at one end of another unit interval and at right angles to it, we are in effect constructing a right-angled isosceles triangle with sides of lengths 1, 1, and $\sqrt{2}$. In particular this shows that the number $\sqrt{2}$ is *constructible*, meaning that, given an interval to act as our standard unit, we can, using straight edge and compasses, construct another interval of length $\sqrt{2}$. The above squaring of a rectangle can be recast in the same language saying that, given two lengths (corresponding to the sides of our rectangle), we can construct the square root of their product (corresponding to the side of our square). If we take one of these sides to have unit length we discover in particular that we can construct the square root of any given number (corresponding to the side of the rectangle not of unit length). In other words, the class of constructible numbers is *closed* under the taking of square roots. It is natural to ask whether the class is similarly closed under the taking of cube roots. In particular, the ancient Athenians asked whether it is possible to construct the cube root of 2? (Legend has it that this was the task set by the god in order to banish the plague from Athens, set in the form of doubling the size, that is to say volume, of a cubic altar.) We shall return to this question with a more systematic approach.

The Simple Constructions

The simplest constructions are very natural—anyone left to play with compasses and edge are likely to discover them for themselves before too long. For example, to bisect a given line AB open your compasses to a radius *at least half that of AB* and draw circles of that radius centred at A and at B, meeting at two points, C and D, one above and the other below AB. The line CD is then the perpendicular bisector of AB and the triangle $\triangle ABC$ is *equilateral* so, in particular, the angle $\angle CAB$ that you have constructed is one of $60°$. Another basic construction is that of a line parallel to a given line L through a specified point, P. This can be done in a variety of ways.

Take any two points A and B on L and, with centre P and radius AB, draw a circle S; with centre B and radius equal to AP describe a second circle to cut S at Q as shown: PQ is then the required line (see Fig. 7.2). Why does this work? The figure $APQB$ is a quadrilateral with $AB = PQ$ and $BQ = AP$. Since opposite sides are equal, they are parallel (a basic result that we shall use more than once), that is to say, the figure is a parallelogram and in particular the line PQ is parallel to AB.

Armed with this construction, we can divide a given line segment AB into an equal number of parts; we describe the case of five parts but any number is possible in the same way. Simply draw another line through A using your straight edge and with compasses mark off any five identical distances along the line starting from A; join the last of these points to B and then construct lines through your marked points parallel to this auxiliary line. The parallels will then meet the original line AB in five equally spaced points and so partition the original line into five equal segments. That this is so follows from the construction as the five triangles involved are similar.

First constructions that come to mind are not necessarily the best. The previous method would have you constructing four parallels and in general, to partition a line into n equal parts would require the construction of $n - 1$ parallels. A second method allows you to do the same job with the construction of only one parallel, thereby radically reducing the complexity of the construction. As before we draw a line AC at any convenient angle to AB and mark off four equal parts along AC as before at points P,Q,R,S. Now construct a line BD parallel to AC as shown in Figure 7.3 and mark off the same four equal intervals on BD, measuring from B at points P', Q', R', S' say. Finally join PP', QQ', etc. to meet AB at p, q, r, s,

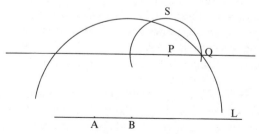

Fig. 7.2: Construction of a parallel line through a given point P

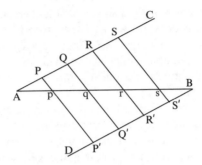

Fig. 7.3: Dividing a given line into any number of equal parts

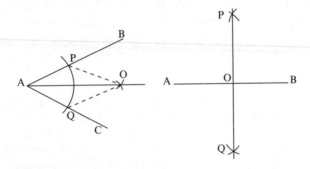

Fig. 7.4: Bisection of a given angle and a given line

thereby dividing the given line *AB* into five equal parts. Again, it is the properties of similar triangles that make all this work: if a line *L* crosses a series of parallel lines and the intercepts of *L* with the parallels are all equally spaced, then the same is true of *any* line crossing the same set of parallels.

Other simple constructions using the Euclidean tools are not hard to find. Figure 7.4 shows bisection of a given angle ∠*BAC* and the bisecting of a given line *AB*.

To bisect the angle ∠*BAC* draw an arc of a circle centred at *A* cutting the sides at *P* and *Q* as shown; then with centres *P* and *Q* and radius *PQ* draw two arcs that will meet at a point *O*. The line *AO* then bisects the original angle. To produce the perpendicular bisector of *AB* take as radius *AB* and draw arcs centred at *A* and *B* respectively that will meet at two points, *P* and *Q*, as shown. The line

PQ is then the perpendicular bisector of AB meeting that line at the point O and, as noted earlier in the chapter, △APB is equilateral.

The pair of constructions in Figure 7.5 show how to erect a perpendicular to a given line AB at a point X on the line and, in the second case, through a point X external to the line. In the first case locate a pair of points, P and Q, on the line equidistant from X (using compasses centred at X) and then, centring the compasses at P and then at Q, draw arcs of common radius exceeding PX so they meet above X at O, yielding the required perpendicular XO. In the second construction we locate P and Q on the line equidistant from X and then bisect PQ, giving the required perpendicular OX as shown.

Figure 7.6 demonstrates how to construct a copy of a given angle ∠BAC at a given point O on a given line FG, while the second

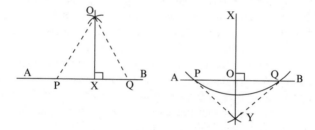

Fig. 7.5: Erection of perpendiculars to a line through a given point X

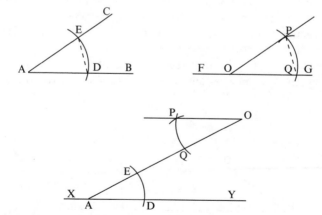

Fig. 7.6: Duplicating an angle and constructing a parallel

picture uses this to provide an alternative way of constructing a line parallel to another, *XY*, and through a given point *O*. Once the general approach becomes familiar, the pictures begin to speak for themselves.

Another use of the parallel line construction is in providing a triangle equal in area to that of a given quadrilateral, *ABCD*. We here revisit our concave quadrilateral of Chapter 6 (see Fig. 6.4). As in Figure 7.7, through *C* construct a line parallel to the diagonal *DB* that meets the extended line *AB* at *X*. The triangle △*ADX* then does the job. The quadrilateral and our triangle each consist of △*ADB* together with another triangle: △*DCB* for the quadrilateral and △*DBX* for △*ADX*. However, these two additional triangles have a common base, *DB*, and equal altitude with respect to this base as *CX* is parallel to *DB* and so the triangles in question share the same area.

The construction is based on the observation that as the vertex *C* of *ABCD* is moved parallel to the diagonal *DB*, the area of △*DCB*, and hence the overall area of the quadrilateral, does not change. Eventually the point *C* meets the line *AB* where the four-sided figure degenerates to a triangle still with the same area. The idea can be extended to polygons with more than four sides. Given an *n*-sided polygon we can cut one corner and move its vertex *C* parallel to the diagonal cut until eventually it meets the extension of a neighbouring side. This will give a polygon with *n* − 1 sides that still has the original area. Repetition of the construction will eventually yield a triangle with area the same as that of the orginal polygon. If we wish we may now easily construct a rectangle, and indeed a square, carrying the same area as the original figure.

It is natural to ask which regular polygons can be drawn under this regime of compasses and straight edge. As we have just seen,

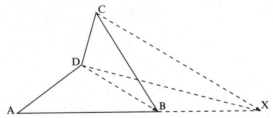

Fig. 7.7: Triangle with same area as a given quadrilateral

equilateral triangles are easy, as are squares. Also it is possible to double the number of edges of a given polygon, that is to say, if a regular n-gon can be constructed, so can a regular $2n$-gon. All we need do is construct the n-gon and find its centre. For a polygon with an even number of sides simply find where the diagonals meet, for an odd number of sides bisect two of the sides and they will meet at the centre (that is, the centroid in the sense of Chapter 4) of the figure. (And this applies to circles as well: the centre is the intersection of the bisectors of any pair of non-parallel chords.) Once we have the centre, C, draw a circle centred at C that passes through one, and hence all, of the vertices of the polygon. Now bisect each edge of the n-gon and where the bisectors meet the circumscribed circle gives you the missing corners of your regular $2n$-gon. (Indeed, only one bisection is necessary: copies of the required side length may then be transferred elsewhere in the diagram as necessary.) In particular we may construct regular hexagons and octagons in this way from equilateral triangles and squares respectively. The more difficult problem of construction of the pentagon we shall return to later.

A number of constructions exploit the circle theorems mentioned in Chapter 3. Our next example, construction of the *direct common tangents* to a given pair of circles, makes use of the fact that the angle in a semicircle is a right angle (recall Fig. 3.13(b)). The task is to find the pair of common tangents that do *not* cross one another for two given circles with centres A and B and respective radii a and b—the completed picture gives the shape of a fan belt connecting two wheels of different radii (see Fig. 7.8).

It does no harm to take a to be at least as large as b, and we draw a circle of radius $a - b$ with centre A. Next we construct a circle with diameter AB; this cuts our first constructed circle at point P. Extend the line AP to meet the larger circle at X, say, and finally construct a line parallel to AX through B that meets the smaller circle on the right at Y, say. The line XY is then tangent to both the original circles and the second direct tangent may be similarly constructed and is the reflection of the tangent XY in the line AB.

Why does this work? First observe that

$$XP = XA - PA = a - (a - b) = a - a + b = b = BY,$$

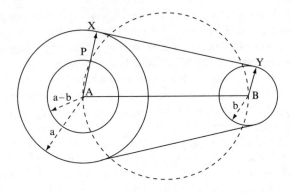

Fig. 7.8: Construction of direct common tangent pair to two given circles

and so *XP* is parallel and equal to *BY*, and this ensures that *XPBY* is a parallelogram. It is in fact a rectangle: $\angle APB$ is a right angle as it is an angle inside the semicircle with diameter *AB*. It follows that $\angle XPB$ is also a right angle, being the supplementary part of the straight angle at *P*, and since opposite angles in a parallelogram are equal, we deduce that all four angles are right angles. In particular, the fact that the angles $\angle AXY$ and $\angle BYX$ are each right angles guarantees that *XY* is tangent to both the original circles.

The special case where $a = b$ in the preceding construction leads to a simplification as one would expect: we merely need to erect perpendiculars to *AB* at *A* and at *B* to find the points where the common direct tangents meet the given circles. In a similar manner, the common *transverse tangents* can be constructed. These tangents go from the upper part of one circle to the lower part of the other and so cross one another at a point along the line joining the two centres of the given circles. You should be able to discover how to proceed based on Figure 7.8: on this occasion you introduce a circle of radius $a + b$ with centre *A* but otherwise things proceed in similar fashion. Indeed, by treating the circle of radius $a - b$ as one of the given circles, Figure 7.8 can easily be modified to give the transverse tangents for the pair of smaller circles in the picture.

Some other constructions along these lines are that of finding a circle through three given points (the centre is found by bisecting a pair of chords determined by the given points), drawing a tangent to a circle from a given point outside of it (this makes use of the semicircle introduced in our previous example), and finding a circle through two given points, A and B, and touching a given circle. For this problem, first construct any circle through A and B large enough to meet the given circle at two points, P and Q; let D be the intersection of the lines AB and PQ and construct a tangent to the given circle from D to meet it at C, say—the circle through A, B, and C is the one required. (The justification for this construction is based on the intersecting secants theorem mentioned earlier in Fig. 3.13(e).)

Constructions Involving the Golden Ratio and Fermat Primes

Although it is easy to construct regular polygons with three, four, and six sides each, constructing the regular pentagon is not quite so easy for, although the constructions for the three former cases come about in the most natural of ways, that of the pentagon is not one you could expect someone to discover by themselves, at least not very quickly. You will recall that when discussing tilings we found that, although three, four, and six-fold rotational symmetry was possible, five-fold symmetry was impossible, at least for a periodic covering of the plane. This perhaps gives us a forewarning that the pentagon may pose some difficulties. However, they can be overcome.

The internal angle of a regular pentagon leads to some special symmetries of the figure (see Fig. 7.9). The obtuse angle between any two sides of the pentagon has value 108° (see pp. 125–6) and this leads to the quadrilateral EPCD being a rhombus, that is to say, a parallelogram with all sides the same length. (The details of this and other claims that follow are not difficult and can be read in the final chapter.) This in turn leads to a very special relationship in that diagonal pairs such as AC and BE in the picture meet one another in the *golden ratio*, which is to say that *EB/EP* is equal to *EP/PB*. Of this symmetry Kepler himself believed that it had 'served the Creator as an idea when He introduced the continuous generation of similar objects from similar objects'.

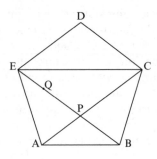

Fig. 7.9: Pentagon and the golden ratio

To illustrate what Kepler had in mind it is easy to show that this particular proportion is self-generating in that, if the distance PB is now marked out from P towards E on the diagonal to the point Q, then $PE/PQ = PQ/QE$ and so we have divided the smaller line EP in the same golden ratio as was the original diagonal. And this process by which the smaller section of the previous division can act as the larger section of the next can be repeated indefinitely, generating a sequence of ever smaller copies of the orginal diagonal divided into two parts related by the golden ratio. Equally we could unfold the process in the opposite direction to create larger copies: extending BE beyond E by a distance EP to a point F we find that E now divides the interval BF in the golden ratio. This idea of self-similarity is the basis of the more modern notion of *fractal* and will emerge again when we look at the Penrose non-periodic tiling later in the chapter.

Let us use the side length of our pentagon to be constructed as our unit of measurement and let d be the golden ratio, the length of a diagonal of the same pentagon. Euclid showed that d is the result of the very simple construction of Figure 7.10 which, at first glance, seems unrelated to the pentagon. We take BC as our unit length and construct the square $ABCD$ as shown. Find the midpoint M of AB and let P be the point where the circle centred at M of radius MD meets the extension of the line BA. The length of the line BP is then d, the golden ratio and the length of the diagonal of our regular pentagon.

The availability of the length d is all that is needed to solve the construction problem of the regular pentagon: given a side of unit

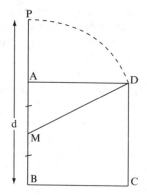

Fig. 7.10: Euclid's construction of the golden ratio

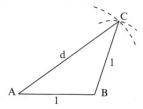

Fig. 7.11: Construction of the regular pentagon

length AB and the length d, the next vertex C of our pentagon in Figure 7.9 is then situated at the intersection of the circle of radius d centred at A and the unit circle centred at B (see Fig. 7.11). Repeating this procedure we then locate the other vertices D and E in turn and so construct our regular pentagon.

Through the construction of the equilateral triangle and successive doubling of side numbers we see that regular polygons with 3, 6, 12, 24, ... sides may be constructed and since we may construct squares and, as we have just seen, regular pentagons also, we know there are constructions for regular polygons with 4, 8, 16, 32, ... and 5, 10, 20, 40, ... sides as well. A little thought reveals more possibilities: we can construct the regular *quindecagon*—the regular polygon with fifteen sides. To see this we can follow the approach used in the pentagonal case and observe that we only need show that it is possible to construct the angle between two adjacent edges of the figure or, what is equivalent, the (acute) exterior angle. The sum of all the exterior angles of a convex polygon is 360° so the exterior angle of the quindecagon is therefore $(360/15)° = 24°$. Now since the exterior angle of a pentagon is 72° while the interior angle of an equilateral triangle is 60° we can subtract the latter from the former to create a 12° angle and by copying this on top of itself we can construct the required angle of 24°. Therefore the regular polygon on fifteen sides can be drawn using the Euclidean tools, as can that on 30, 60, 120, ... sides.

The question of exactly which regular polygons can and cannot be constructed using straight edge and compasses alone was settled by Carl Frederich Gauss (1777–1855) at the age of 19. As we have already noted, given that a regular polygon with n sides can be constructed we can, by bisecting each side, construct a regular polygon with $2n$ sides. What Gauss proved was that a regular polygon with n sides can be constructed with Euclidean tools if and only if n is a product of the form

$$n = 2^m p_1 p_2 \cdots p_k;$$

where the factors p_i are distinct *Fermat primes*. The reasons why powers of 2 should feature in the solution has already been explained but what are these Fermat primes?

The nth *Fermat number* F_n is

$$F_n = 2^{2^n} + 1, n = 0, 1, 2, \ldots$$

For the first few values of n these numbers are prime:

$$F_0 = 3, F_1 = 5, F_2 = 17, F_3 = 257, F_4 = 65\,537,$$

and so, according to Gauss, it is possible to construct regular polygons with the corresponding number of sides: we have seen how to construct the F_0-gon and the F_1-gon and Gauss himself underlined this point by explicitly constructing a regular 17-gon (the F_2-gon). The construction of the regular quindecagon is also consistent with this result as $15 = 3 \times 5$, the product of the first two Fermat primes ($F_0 = 3$ while $F_1 = 5$). The feat of constructing a regular $F_3 = 257$-gon was accomplished by Richelot and Schwendenwein in 1832 and a certain J. Hermes spent ten years on the 65 537-gon and deposited his effort in a box at the University of Gottingen where it may still lie.

What of the higher Fermat numbers? Fermat himself claimed that the next Fermat number, $F_5 = 4\,294\,967\,297$, was prime and conjectured that they all were. The Fermat primes are indeed peculiar, not only because of their surprising link to geometry but because they have little right to be prime at all: a number of the form $a^n - 1$ is never prime unless $a = 2$ and n is itself prime (and then not always); $a^n + 1$ can only be prime if a is even and n is a power of 2. Therefore we see that for $a = 2$ the only prime candidates remaining of this form are the Fermat numbers, F_n. (For detail see Chapter 8.) On the face of it, F_5 looks likely to be prime as it certainly fails the divisibility tests for all small primes so it came as quite a surprise when, in 1732, nearly a century after Fermat's lifetime, Leonhard Euler noticed that F_5 is divisible by 641 and so the conjecture was false:

$$F_5 = 4\,294\,967\,297 = 641 \times 6\,700\,417.$$

In 1880 it was found that F_6 has $274\,177$ as a prime factor and since then *every* Fermat number whose primality question has been settled has been found *not* to be prime: the present list goes at least as far as F_{16} so it is safe to say that the effort of Hermes in constructing the F_4-gon will never be surpassed!

How did Euler do it? Of that we cannot be sure but, once this arithmetical rabbit has been pulled out of the hat, anyone can check the above multiplication in a few minutes. However, if you are familiar with modular arithmetic, you will find in Chapter 8 how that technique, allied with a little numerical detective work, can show that 641 is a factor of F_5 without actually performing the long division.

Breaking the Rules

Frustration with the apparent limitations of the Euclidean tools led the Greeks to all manner of devices and curves for solving the apparently insoluble trio of Delian problems, those being the *Duplication of the Cube*, that is, the construction of the cube root of 2, the *Trisection of the Angle*, and the *Squaring of the Circle*. In this they had some success but the Delian problems were never resolved satisfactorily, partly because the new methods involving various higher curves and other mechanical devices smacked of the arbitrary but more fundamentally because the nagging doubt persisted: was it possible these problems could be solved with straight edge and compasses but no one had been clever enough to find out how?

As will be explained to a greater extent later, the settling of the Delian problems in the negative would have to await the development of algebra quite beyond anything in the mathematics of antiquity, especially in the case of the Squaring of the Circle whose impossibility was not decided until 1882. On the other hand, some modern European mathematicians also were keen to explore the opposite direction and look for minimalist approaches, asking whether we could restrict the Euclidean tools further without the loss of theoretical constructive power. The most well-known results along these lines were published in 1797 by the Italian Lorenzo Mascheroni in his *Geometria del compasso* in which he established the surprising conclusion that any Euclidean construction could be carried out using compasses alone! Of course, a straight line cannot be *drawn* without an edge but Mascheroni showed that any constructible line could be *determined* using only his compasses by locating two points on it. In a curious quirk of history it turned out that this had been proved 125 years earlier by an obscure Danish mathematician, Georg Mohr, in a book he published in 1672 entitled *Euclides Danicus*. The discovery went totally unnoticed until the work turned up in a Copenhagen bookshop in 1928, to the delight and astonishment of the buyer, the Danish mathematician Hjelmslev.

As early as the 10th century the Arabian mathematician Abu'l-Wefa considered constructions using a straight edge and 'rusty' compasses that could only draw circles of a single fixed radius. In 1833, in the wake of Mascheroni's result, Poncelot and Steiner showed that not all Euclidean constructions were possible with an edge alone but everything was possible with an edge and a *single* circle with its centre marked on the page; in particular Abu'l-Wefa's rusty compasses were enough to do Euclidean geometry. These results now seem to have been pushed to their limits in two directions. In 1904, the Italian Francesco Severi proved that, along with the edge, we do not need the whole of the Steiner circle but only require an arc, however small, and the position of the centre to carry out any Euclidean procedure, while at the other extreme we can do away with compasses entirely as long as we have a *two-edged* straight edge, whether or not the edges are parallel. (Again, we could not *draw* a circle without compasses, but we could specify it by three points on the circumference.)

Although it may be possible to lead a very austere geometrical life with such rudimentary tools there is a price to be paid, in that the number of operations involved to carry out a construction increases tremendously if we insist on handicapping ourselves in these ways and that is the other side of the coin. Studies have also been made of how to perform Euclidean constructions *efficiently*, looking to minimize the total number of operations carried out (in order to do things quickly) and the total number of intersections of lines and circles involved (in order to do things accurately). These general questions are extensively pursued throughout that part of modern mathematics involving algorithmic procedures, that is to say, those, such as are used in computing, for executing specific tasks. On the one hand, mathematicians want to learn the precise theoretical limits to tasks possible under a certain regime of action and, on the other, they are equally concerned with what can be done in real time.

Let us look at a sample of Greek tricks for dealing with the Delian problems. Since it is so easy to trisect a line (or indeed divide it into any number of equal parts as shown in Figure 7.3) it seemed surprising that trisection of angles should be so difficult. Of course it is possible to trisect *some* angles, for example we can bisect 60° to give us 30° and thereby trisect a right angle but the Greeks could find no way to trisect 60°, which is what is required to constuct the regular *nonagon*, the nine-sided polygon. Why should it be that we can trisect any length and some angles, but not others? This was a question they could not answer but Archimedes showed that it was quite simple to trisect any angle through the use of a spiral that connected linear and angular velocities that then allowed him to transfer line trisection into angle trisection.

Consider the path of a point P that moves at a constant speed away from a fixed origin O in such a way that the angular speed of the line OP is also constant. Its path traces out what is appropriately called an Archimedean spiral as pictured in Figure 7.12. Let $\angle AOB$ be

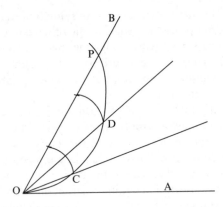

Fig. 7.12: Trisection of an angle with an Archimedean spiral

the angle to be trisected and let *P* be the point where the spiral meets the arm *OB* of the angle. Trisect the *line OP* and draw arcs of circles, each centred at *O* through the trisection points to meet the spiral in the points *C* and *D* as shown. Since the angle ∠*AOP* is directly proportional to the distance of *P* from *O*, the angle ∠*AOC* and for that matter ∠*COD* are each equal to one third that of the given angle ∠*AOB*. In this way Archimedes trisected the angle by use of his spiral.

It can also be shown that it is possible to square the circle if such free use of the spiral is permitted. Squaring the circle proved to be a particularly tantalizing problem for as early as the 5th century BC Hippocrates had shown that it was possible to find the exact area of certain *lunes*, crescent-shaped areas between two circles of differing radii, and it was realized that if something similar could be done for *any* lune shape then it would be possible to square the circle.

As was mentioned in Chapter 5, the curves arising from conic sections also give rise to intersections that can be interpreted as solving the duplication of the cube and that is why the curves were originally introduced by Menaechmus around 350 BC. Another solution due to Archytas (*c.*400 BC) rests on locating the common point of a cylinder, a torus of zero inner diameter, and a cone. Other curves generated in a mechanical manner also lead to angle trisections such as the bell-shaped *Conchoid of Nicomedes* (*c.*240 BC).

It was also appreciated that certain figures could resolve one or other of the Delian problems, although they could not be drawn with Euclidean tools alone. For example, the simple diagram in Figure 7.13 involves a line of length $\sqrt[3]{2}$.

Consider the two right triangles $\triangle DAB$ and $\triangle ABC$ where AC and BD meet at right angles, PC is taken to be our unit length and $PD = 2$. By looking at angles involved we see that both our triangles are similar to each other and to the smaller right-angled triangles $\triangle BPC$, $\triangle APB$ and $\triangle DPA$. Given all this, comparison of sides of these similar triangles reveals the relationships

$$\frac{x}{1} = \frac{y}{x} = \frac{2}{y},$$

which in turn yield the equations $x^2 = y$ and $y^2 = 2x$: squaring the first and using the second then gives $x^4 = y^2 = 2x$, whereupon cancelling the common factor x yields $x^3 = 2$, so that x is the cube root of 2.

To construct this figure we may begin by taking D anywhere sufficiently above A so that we can rotate the L-shape, DPC, about D until the extension of CP passes through the point A; then by adjusting D either up or down and allowing PA to contract or extend as required we could ensure that C lay directly above B, the point where the extension of DP meets the horizontal. Indeed this

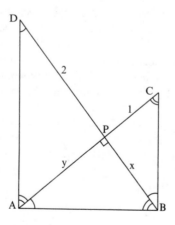

Fig. 7.13: Constructing the cube root of 2

figure can be constructed if we allow ourselves free use of the traditional draughtsman's tools of carpenter's square and set square.

It is also a simple enough matter to trisect an angle if we allow ourselves to mark a given line segment not just on the paper but also on our straight edge for then we can produce the situation pictured in Figure 7.14. Here we are given an angle $\angle ABC$ and we construct the rectangle $ADBC$ with diagonal AB as shown. Contrary to the Euclidean rules we then mark the distance $E'F'$ on our edge of length $2BA$. We then place E' along AC and by placing a peg at B for the edge to act against we ensure that the edge passes through B; we then move the point E' up or down the line AC while keeping the edge running through B until the point F' lies on the extension of the line DA. The positions of E' and F' are now marked as E and F respectively. Having done this we find that the angle $\angle EBC$ is one third that of the given angle $\angle ABC$. To see this, let G be the midpoint of EF. Since G is the centre of the rectangle with sides EA, AF and $BA = \frac{1}{2}EF$ we see that

$$EG = GF = GA = BA,$$

whereupon we get the following equality of angles;

$$\angle ABG = \angle AGB = \angle GAF + \angle GFA = 2\angle GFA = 2\angle GBC,$$

so that $\angle ABC = \angle ABG + \angle GBC = 3\angle GBC = 3\angle FBC$, that is to say the line BF trisects the angle $\angle ABC$.

The three angle bisectors of a triangle meet at the one point, the incentre (see Fig. 4.11), but what of the angle trisectors? This simple question has an equally simple answer: adjacent trisectors meet in pairs forming an equilateral triangle, yet this elementary result is not to be found in the books of Euclid as it was only discovered by F. Morley in 1899 (Fig. 7.15). To be fair, the proof is tricky (see

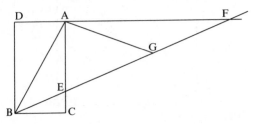

Fig. 7.14: Trisection of the angle $\angle ABC$

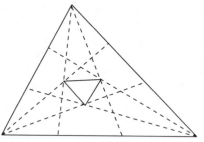

Fig. 7.15: Adjacent angle trisectors form an equilateral triangle

Chapter 8) but one explanation for the mathematical community overlooking so elementary a fact for so long (about 2 400 years) is that angle trisection was in such bad odour no one dared think seriously about it. Although the trisection is impossible using Euclidean tools, the trisectors certainly exist and can be constructed by other means, so there is no reason not to study them and reveal any special properties they have. It could be that mathematicians, like everyone else, can be adversely influenced by fashion, and so grow inhibited and timid—never a good thing in science.

Describing the Limits of Euclid's Tools

To explain, for instance, why it is impossible to construct the cube root of 2 with Euclidean tools requires an algebraic language not available until the 19th century that can never the less be outlined here. The first attempt at this line of argument seems to have been hinted at much earlier in the work of Leonardo of Pisa.[1] In 1220 he conducted an argument whose purpose was to show that a certain cubic equation could have no solution of the form

$$\sqrt{a} + \sqrt{b},$$

which leads to the conclusion that a root cannot be constructed with straight edge and compasses.

---

1. Better known as Fibonacci, the man who introduced the famous Fibonacci sequence of numbers, 1, 1, 2, 3, 5, 8, 13, 21, . . . that turns out to be intimately connected with the golden ratio. Each number after the second is the sum of its two predecessors.

Let us begin by recapitulating what is possible. Given two lengths, *a* and *b*, we can, through use of the Euclidean tools, construct lines corresponding to their sum, difference, product, quotient, and square root. Indeed the four arithmetical operations require no more than construction of certain parallels and the resultant similar right-angled triangles but to take the square root we need a special case of the construction for squaring of the rectangle that opened the chapter. All this is summarized in the pictures of Figure 7.16.

We may therefore, beginning with a unit length, use the Euclidean tools to produce line segments of any rational length, that is to say lengths of the form *a/b*, where *a* and *b* are positive or negative whole numbers or *integers* as they are called. The set of all rational numbers, *F*, forms what is known as a *field*, meaning that the collection is closed under the four arithmetical operations: performing arithmetic with rational numbers always leads to answers that are themselves rational numbers. We can however escape from this field by taking roots and in particular we can construct the square root of 2 that, as we saw in Chapter 3, does not lie within the field of the rationals.

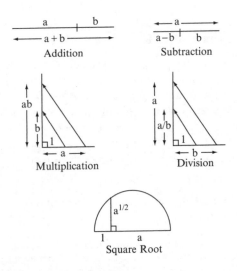

Fig. 7.16: The four arithmetical operations and square roots using Euclid

For any positive rational c we can now construct any length of the form $a + b\sqrt{c}$, where a and b are rationals and this larger collection of numbers also forms a field, F_1 (that depends on our choice of the number c).[2] Next we may take any positive number d in our new field, F_1, and construct any number of the form $a + b\sqrt{d}$, where a and b come from F_1 also. This leads to a new and generally larger field of constructible numbers F_2, and so on: we can construct a chain of larger and larger fields of constructible numbers in this way.

What is more, it is not especially difficult to verify that every constructible number arises in this manner. That is to say the set of all constructible numbers is exactly the set of numbers that can be got by beginning with integers and performing the operations of addition, subtraction, multiplication, division, and the taking of square roots as many times as we please. This is because new lengths come about in Euclidean constructions through intersections of lines and circles of given specifications and the equations describing lines and circles are linear in the case of lines and a simple type of second degree equation in the case of circles.

We can therefore construct lengths like

$$\frac{3}{13} + \sqrt{\sqrt{61} - 7\sqrt{2 - \sqrt{3}}}$$

but not a length like $\sqrt[3]{2}$ as that cannot be written in this form. However the potentially embarrassing question still remains: how do we *know* that it is not possible to write $\sqrt[3]{2}$ as some (perhaps very complicated) expression involving only rational numbers and square roots. Our intuition may suggest that we will not be able to manufacture a cube root out of square roots but it is not enough to argue that, if it could be done, surely it would not be too difficult to see how. If this were the best we could do we would have made no real progress at all as we would be just as entitled to make the same claim for the problem when viewed in its original setting: if we could construct cube roots using Euclidean tools we would have found out how to do it by now. Happily, this algebraic formulation does grant us a sufficiently strong hold on the question to settle it in the

---

2. It is not obvious that we can divide one number of this type by another and simplify the answer to a third number of the same form which is what is required for F_1 to be closed under division, but we can (see Chapter 8).

negative: the description provided here is precise enough to allow an argument that shows that the cube root of 2 cannot arise in this fashion and therefore is *not* a constructible number. Further detail is given in the final chapter.

Tilings and the Golden Ratio

We shall close with some words about tilings that are not periodic and the connection and relevance comes via the golden ratio, although there is much more to the story than can be explained here. All the tilings of Alhambra of the previous chapter were periodic tilings that allowed translational symmetry in two directions—the entire plane could be partitioned into copies of a certain fundamental region that repeated itself in all directions. It is possible to tile the plane with a fixed number of tiles in a fashion that is not repetitive in this way. Indeed it can be done with a single tile.

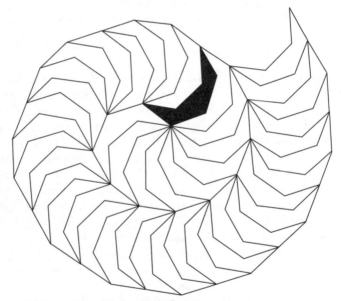

Fig. 7.17(a): A nonperiodic tiling by a single tile type

An example, due to B. Grunbaum and G. C. Shephard, uses their so-called *versatile* which is a heptagon, a seven-sided polygon, that can cover the plane in a spiral pattern, one fragment of which is shown in Figure 7.17(*a*). Each heptagon nestles into the next with a twist, although the shaded versatile is twisted clockwise compared to its predecessor while others are twisted anticlockwise. Since the centre of the spiral is unique, the tiling is not periodic, that is to say does not have translational symmetry. The centre may act as our origin or point of reference from which we could measure coordinates and it is a recognizable point. This contrasts with our periodic tilings where, for any point *P*, there are infinitely many points in the plane from which the overall view as we look out is exactly the same as the way it appears from *P*.

The same tile can however cover the plane in a tame periodic pattern. We can in fact lay down copies of the tile to cover parallel strips by nestling the tiles into one another using alternate clockwise and anti-clockwise twists and so the versatile also admits a periodic tiling of the plane (see Fig. 7.17(b)).

The question then arises whether or not it is possible to find a finite set of tiles that cover the plane but *only* in an *aperiodic* (that is

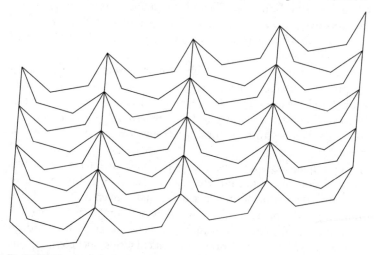

Fig. 7.17(b): The same tile covering the plane periodically

nonperiodic) fashion. The significance of this question came to light via another mathematical avenue that I must take a moment to explain. During the first half of the 20th century it was found, much to everyone's initial surprise, that there are some mathematical questions for which it is not possible to write down a set of instructions that will allow you to answer the question. For example, consider the general problem of determining whether one number n is a factor of another number m. Since there are infinitely many numbers, there are infinitely many instances of this problem. However, it *is* possible to describe a general procedure (long division) which, when applied to *any* problem of this type, will yield the answer, yes or no. The mathematical word for such a procedure is an *algorithm*, a finite list of instructions, capable of translation into a computer programme, that will solve all instances of your problem. However, it was proved from the work of Kurt Godel, Alan Turing, and others that mathematics truly is hard in that there are similar questions that are *undecidable*, meaning that there is no such mechanical procedure for dealing with them. A typical source of intractable examples is so-called *word problems*, where we are interested in whether one string of symbols can be transformed to another using various substitution rules. There is no single procedure that allows you to solve every such problem of this kind although some special kinds of word problems may be solvable and indeed may be solved very readily.

Some of the first undecidable problems were deliberately constructed by mathematicians simply to show that such horrors existed and the questions asked were not necessarily of any intrinsic interest. However the warning had been sounded and it was not too long before some natural problems turned out to be undecidable.

One question along these lines presently of interest to us was asked by Hao Wang in 1961. He posed a question that is equivalent to asking whether or not it is decidable that a given finite set of tiles could cover the plane. In other words, what Wang was seeking was some precise procedure by which you take the given tiles and go through a well-described process of calculations and comparisons that is guaranteed to eventually finish with the answer to the question resolved one way or the other. The answer he found was at the time inconclusive: he showed that there was such a procedure so long as it were true that whenever a finite set of distinct tiles could cover the plane then it was possible to tile the plane with those tiles in a periodic fashion. Objects like the versatile that can tile the plane both in periodic and nonperiodic ways did not cause trouble, only the existence of tile sets that covered the plane exclusively in aperiodic ways were problematic.

It would seem that up until this time no one had seriously considered the possibility that such strange tiling sets may exist but Wang's result stimulated study of the question and led to the surprising announcement in 1966 by Robert Berger of the discovery of a set of 20 426 tiles that could tile the plane but only without any translational symmetry. The example seemed horrendously complicated but none the less its very existence thwarted the possibility of there ever being a tiling algorithm: Wang's tiling problem was undecidable—we know for certain that there is no computer programme that will always be able to tell you whether a given set of tiles will cover the plane. This question is at a higher level of difficulty than can ever be dealt with by mere calculation.

By further analysis Berger was able to reduce his huge counterexample to one involving only 104 tiles and in 1971 an example involving a set of just six tiles was devised by Raphael Robinson. By 1973 Roger Penrose had produced examples involving just a pair of tiles that covered the plane only aperiodically. It remains an open and very intriguing question whether there is a single shape that covers the plane but only in an aperiodic fashion.

An example of a pair of Penrose tiles is shown in Figure 7.18. The underlying pair of tiles of this Penrose tiling are on the right in Figure 7.18 and are referred to as the thick and thin rhombs. Each is based on the pentagonal angle of $\theta = 36°$ (the angle separating adjacent diagonals of the regular pentagon). The angles between sides of the thin rhomb are θ and 4θ while within the thick rhomb we see angles of 2θ and 3θ. Given these angles, we should not be surprised to see the golden ratio emerging: if we assign unit length to the sides of each rhomb, then the length of the long diagonal of the thick rhomb T is d, the golden ratio, while the length of the short diagonal of the thin rhomb t is $1/d$, the reciprocal of the golden ratio. Deforming the edges of the thin and thick rhombs we obtain the

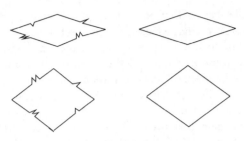

Fig. 7.18: The thick and thin rhomb Penrose tiles

tiles on the left that are the actual Penrose tiles. The purpose of these edge deformations is to impose certain matching rules between the tilings—obviously the protruding single and double teeth are there to match the corresponding gaps and these additional features are designed to forbid certain edge-to-edge matchings that lead to ordinary periodic coverings of the plane. It is easiest to think of the Penrose tiles in terms of the underlying rhombs, together with the matching rules that are represented by the edge deformations. A fragment of a (necessarily aperiodic) tiling of the plane by the Penrose rhombs is to be seen in Figure 7.19.

The tilings of Penrose almost have translational symmetry and, at the same time, almost have five-fold rotational symmetry. As we saw in Chapter 6, translational symmetry and five-fold rotational symmetry are incompatible in a tiling but here we come as close as possible to violating that rule in that, in a sense that can be made precise, the pattern goes on to itself through rotation and a turning of one fifth of a circle to any pre-assigned degree of accuracy just short of absolute perfection. In this way we have a phenomenon akin to the (real) quasicrystals discovered in the 1980s. Without quite having translational symmetry, any covering by Penrose tiles is *repetitive*, meaning that any *patch P*, as a region of the tiling is called, can be found over and over again wherever you look throughout the patchwork. Even more is certain: there is a number r (whose value depends on your patch P) with the property that P may be found somewhere within *every* circle of radius r (or larger) drawn on a pavement tiled by the Penrose tilings.

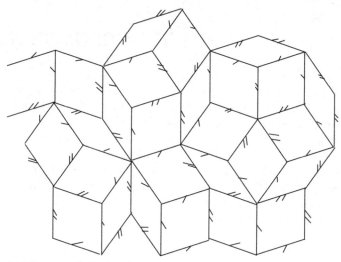

Fig. 7.19: Fragment of a Penrose tiling

Both these phenomena, Penrose tilings and quasicrystals, are not mere novelties but contain real surprises as to their form and structure. The feature they have in common is that their assembly cannot be described at the local level. That is to say, when playing with Penrose tiles and attempting to cover the plane you are likely to get stuck. Even though you have not violated any matching rules in consticting your patch, you may find that the fragment cannot be continued and you have to retreat and partially dismantle your patch before continuing. A corresponding difficulty arises when we consider the mechanism by which a quasicrystal can actually form as it would seem that the atoms need to act as if they are taking account of the behaviour in other parts of the structure with which the atom has no contact. Penrose himself speculates in his book *The Emperor's New Mind* (1989) that some kind of mechanism is active at the quantum level. In any case, how quasicrystals come into existence is still an unresolved problem fraught with real difficulty, which partly accounts for the extreme scepticism with which their discovery was initially greeted.

○ For Connoisseurs

This final chapter gives some of the mathematical details that do not appear in the body of the text. Repetition of core school and university mathematics is not the purpose of a book like this one but some of these details are interesting and may not be quite so easy to find in a standard textbook, so they are outlined here for the benefit of readers who want a hard look. This is not meant to be only for experts though. There is some variability in the level of difficulty of the explanation and any interested reader should be able to gain by sampling from this chapter.

Chapter 3

Leonardo's and Euclid's Proof of Pythagoras' Theorem (pp. 54–5)

Two areas A and B are said to be *congruent by addition* if they can be dissected into corresponding pairs of identical pieces. They are *congruent by subtraction* if corresponding pairs of identical pieces can be adjoined to A and B to give two new figures which are congruent by addition. Here in Figure 8.1 is a diagram that is the essence of a proof of Pythagoras due to Leonardo Da Vinci showing that the square on

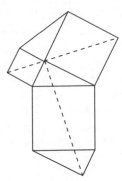

Fig. 8.1: Leonardo Da Vinci's proof of Pythagoras' Theorem

the hypotenuse is congruent by subtraction to the combined squares on the legs of the right-angled triangle. You should be able to convince yourself from the picture that a certain figure can be formed from the large square together with two copies of the triangle and equally with the two smaller squares and two copies of the same triangle. It certainly does no harm to draw, cut, and rearrange the figure to see the proof in action as Leonardo may well have done. As always, bear in mind what would go wrong if the triangle did not have a perfect right angle in the corner.

To recapture Euclid's *Bride's Chair* proof from Figure 3.11 on page 55 first note that by the similarity of the triangles involved we have

$$\frac{DC}{AC} = \frac{AC}{BC};$$

which yields $(DC)(BC) = (AC)^2$. This amounts to saying that the area of the smaller rectangle in the square on the side BC equals that of the square on AC. By the same argument, the square on AB has area equal to that of the remaining rectangle in the square along BC from which it follows that the sum of the areas of the two smaller squares equals that of the larger.

Irrationality of Square Roots (p. 56)

In general the square root of any number is irrational unless that number is a perfect square. The Greeks' geometrical methods did not lend to proving this result and great effort was expended on individual cases. Theodorus of Cyrene around 425 BC managed to prove that all numbers up to 17 that are not perfect squares have irrational roots.[1] A little school algebra is however all you need in order to show that $\sqrt{n}$ is irrational unless n is a perfect square. What we develop below is itself only a special case of the so-called *Rational Root Theorem*.

---

1. It is not clear what method he used and why he apparently had trouble going beyond this point. An informed discussion of this can be read in *An Introduction to the Theory of Numbers* by Hardy and Littlewood (Oxford: OUP, 1938). It may also be noted that the 'Aristotlean' proof of the irrationality of $\sqrt{2}$ works for any prime number p if we make use of Euclid's Lemma that says if a prime p is a factor of a product ab then it is a factor of at least one of the numbers a or b.

Take any polynomial with integer coefficients that is *monic*, that is to say, the leading coefficient is 1: $a_0 + a_1 x + a_2 x^2 + \cdots + a_{n-1}x^{n-1} + x^n$ and suppose that the fraction $\frac{a}{b}$ is a root of that polynomial, that is the polynomial returns the value 0 when $\frac{a}{b}$ is substituted for x. By multiplying top and bottom by -1 if necessary we may assume that b is positive. We can also assume to have cancelled any common factors that $\frac{a}{b}$ may have so that this fraction is in lowest terms. Substitute this root into the polynomial and multiply through by b^{n-1} to gain the conclusion:

$$a_0 + a_1 a b^{n-2} + a_2 a^2 b^{n-3} + \cdots + a_{n-1}a^{n-1} + \frac{a^n}{b} = 0;$$

now the final term on the left must be a whole number because every other term in the equation is a whole number. However, since a and b had no factor in common this can only be possible if $b = 1$. We conclude that *if* a monic polynomial with integral coefficients has a rational root then that root must be a whole number.

We now let n be any positive whole number and suppose that $x = \sqrt{n}$ is rational. We know that $x^2 - n = 0$ so, by the conclusion of the previous paragraph, x is a whole number. Therefore $\sqrt{n}$ cannot be rational unless $n = x^2$ is the square of a whole number, that is to say, a perfect square such as 1, 4, 9, 16, etc. Theodorus would have no doubt have appreciated the power of secondary-school algebra!

John Conway and Richard Guy's *Book of Numbers* has a variety of proofs of the irrationality of $\sqrt{2}$ which they describe as Pythagoras' Big Theorem. Here is a novel one. Suppose that N is a whole number but that $\sqrt{N}$ is not although it is a fraction, B/A, cancelled to lowest terms. Then

$$\frac{B}{A} = \frac{NA}{B}$$

so that the fractional parts of B/A and NA/B have the form a/A and b/B, where a,b are positive numbers smaller than A, B. (They are *positive*, and so not 0, because A is *not* a factor of B.) Since the two numbers in question are equal, so are their fractional parts, giving:

$$\frac{a}{A} = \frac{b}{B},$$

and so

$$\frac{b}{a} = \frac{B}{A} = \sqrt{N}.$$

This gives a simpler form for $\sqrt{N}$, contrary to our assumption that $\sqrt{N}$ was already in lowest form. Therefore if $\sqrt{N}$ is not an integer, then it is not a fraction either.

Proofs of Circle Theorems (pp. 65–7)

The first circle theorem says that two angles standing on the same arc of a circle as in Figure 8.2(a) are equal and have half the value of the angle at the centre. To see that it is indeed the case that $a = 2b$ join the point labelled B to the centre and consider the triangles triangles $\triangle ABO$ and $\triangle CBO$. These triangles are isosceles as each has two sides that are radii of the circle which entitles us to label the unknown angles x and y as shown. Summing the angles around the centre O now gives us in degrees:

$$360 = (180 - 2x) + (180 - 2y) + a,$$

which gives $a = 2x + 2y = 2(x + y) = 2b$, as required. In the case where the angle a is reflex, the theorem still applies and the diagram on the right is the relevant one. By the same type of reasoning we get

$$2x + 2y + (360 - a) = 360,$$

which gives $a = 2x + 2y = 2(x + y) = 2b$ as before, where b is the angle on the circumference at the point B. If we now take a to be a straight angle we gain the corollary that the angle b in a semicircle is a right angle (Fig. 8.2(b)).

The theorem represented by Figure 8.2(c) is still taught in schools but manages to remain obscure. It is written in words as *The angle subtended by the tangent to the chord is equal to the angle subtended at the circumference in the alternate segment.*

This still stands in need of explanation: our chord (of the circle) is CA so that the angle subtended by the tangent and chord is $\angle CAT$ and the claim is that this equals the angle $\angle ABC$ standing on the chord AC but in the segment on the opposite side of the chord to the point T on the tangent that we have used to define our angle. (If we had taken our point T on the tangent to the *left* of the circle then $\angle CAT$ would be the supplementary obtuse angle and the relevant

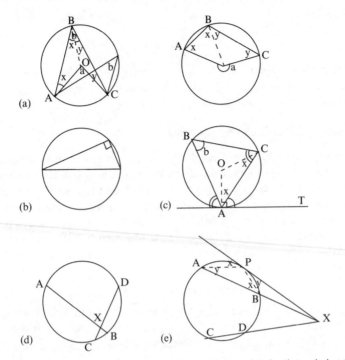

Fig. 8.2: Sample circle theorems. (a) The angle *a* at the centre is twice the angle *b* at the circumference. (b) Taking *a* to be a straight angle, in (a) gives that the angle in a semicircle is a right angle. (c) If *T* is tangent to the circle we have equality of the angles as indicated. (d) Intersecting chords theorem: (AX)(XB)=(CX)(XD). (e) Intersecting secants theorem: (AX)(BX)=(CX)(DX)=(PX)².

angle at the circumference would have been one in which the point *B* is chosen to lie on the circumference in the smaller segment of the circle defined by the chord *AC*.) The point *B* can, of course, be taken anywhere on the circumference in this segment but, by the circle theorem of part (a), the value of the angle ∠*ABC* is independent of where *B* is placed as the angle ∠*ABC* will always equal half the angle ∠*AOC* subtended at the centre. Having clarified the meaning, the proof itself is short: since triangle △*AOC* is isosceles we may label the two angles shown by the same letter *x*. Again calling the angle at *B* by the name *b* we have, on applying part (a),

$$b = \frac{1}{2}(180 - 2x) = 90 - x = \angle CAT.$$

This proves the result which can then be applied to the acute angle that the chord AB makes with the tangent, for it will then equal the angle marked at the point C on the circle.

Next, the intersecting chords theorem (see Fig. 8.2(d)) is based on the observation that the triangles $\triangle AXC$ and $\triangle DXB$ are similar: angles $\angle AXC$ and $\angle DXB$ are a pair of equal opposite angles, while angles $CAX = CAB = CDB = XDB$; the second of the three equalities being true as the angles in question stand on the same arc, CB. Since these triangles share two pairs of angles with the same value they must share three pairs with the same value as the angles in any triangle sum to $180°$. In similar triangles ratios of lengths of corresponding pairs of sides are equal, which gives us in this instance that

$$\frac{AX}{XC} = \frac{DX}{XB},$$

and rearranging this gives the required equality: $(AX)(XB) = (DX)(XC)$.

The final sample result is the intersecting secants theorem: here we are to show that for any chord AB of the circle extended to a point X outside the circle, the product $(AX)(BX)$ equals $(PX)^2$, where P is a point where a tangent drawn from X meets the circle. Given that this is true and applying it to a second secant CDX we get that $(CX)(DX)$ is also equal to $(AX)(BX)$, as each is equal to the square of the length of the tangent. We should note that, as the secant AX rotates about X, clockwise in this case, it eventually approaches the tangent PX and that, in the limit, both of the terms AX and BX approach PX, so that the term $(PX)^2$ is the tangent analogue of the product $(AX)(BX)$ of the secant.

To prove the theorem we can begin by noting that the equality of angle pairs labelled by x and y respectively indicated in Figure 8.2(e) is justified by the alternate angles theorem of Figure 8.2(c) applied to the tangent line PX. Next we note that $\angle APX = 180° - x = \angle PBX$. Since the triangles $\triangle APX$ and $\triangle PBX$ then have two pairs of equal angles ($\angle PXB$ is common to both) they have three pairs equal and so are similar.

Comparing corresponding pairs of sides we get

$$\frac{AX}{PX} = \frac{PX}{BX}$$

whereupon, cross multiplication yields the required equality (AX) $(BX) = (PX)^2$.

The sum of the angles in any quadrilateral is $360°$; another circle theorem that is quite simple to demonstrate is that if the vertices of the quadrilateral happen all to lie on the circumference of a circle then opposite angles in the quadrilateral are supplementary, that is, they sum to $180°$.

Chapter 4

The Sphere and the Cylinder (pp. 72–3)

Archimedes' result on the area A of the cap of the sphere can be obtained quite readily using calculus. Consider the cross-section of the sphere of radius r, where the angle θ and the length y are measured as shown in Figure 8.3. For a given angle θ, the shaded increment of area dA on the sphere is $(2\pi r \sin \theta)(r d\theta)$ and we also have $y = r(1 - \cos \theta)$ so that $dy/d\theta = r \sin \theta$, whereupon, using the chain rule we obtain:

$$\frac{dA}{dy} = \frac{dA}{d\theta} \cdot \frac{d\theta}{dy} = 2\pi r^2 \sin \theta \cdot \frac{1}{r \sin \theta} = 2\pi r.$$

Noting that both A and y equal zero when θ does likewise gives us

$$A = 2\pi r y;$$

and so the area of the cap equals the area of the projection of the cap on to its containing cylinder, as Archimedes told us. In particular, taking $\theta = 180°$ gives the full area of the sphere as $2\pi r(2r) = 4\pi r^2$.

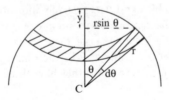

Fig. 8.3: Surface area of a sphere

Archimedes did not have calculus and needed to work much harder: he manipulated certain trigonometric sums that correspond to integrating the sine function from first principles.

Volume of a Spherical Cap

Calculating the *volume* of a cap of a sphere of height a and the radius of whose base is r can now be carried out without further recourse to calculus. The first step is to find the volume of the three-dimensional analogue of the sector of the circle consisting of two parts: the cap of the sphere and the cone whose base is the circular base of the cap and whose apex is the centre of the sphere: this entire object resembles an inverted (headless) toy wobbly clown that always returns to its upright position after being struck. Just like the area of a sector of a circle is directly proportional to the length of the arc defining it, the volume of our 'clown' is directly proportional to the area of the cap defining him which in turn, by the note on the previous page, is proportional to the height of the cap a. Let R denote the radius of the sphere; the fraction of the volume represented by the toy clown is $a/2R$ of the volume of the sphere. From this we need only subtract the volume of the cone to recover the volume, V_c, of the cap:

$$V_c = \frac{a}{2R} \times \frac{4}{3}\pi R^3 - \frac{1}{3}\pi(R - a)r^2.$$

Finally, by Pythagoras, $R, r,$ and a are related by $(R - a)^2 + r^2 = R^2$ which gives $R = (a^2 + r^2)/2a$. Replacing R throughout by this expression and carrying out some more algebraic simplification

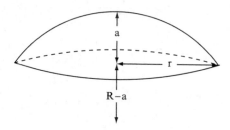

Fig. 8.4: Volume of a spherical cap

yields the formula for the volume of the cap in terms of a and r alone:

$$V_c = \frac{\pi a}{6}(a^2 + 3r^2).$$

Bored Sphere Problem

The cap formula is needed to solve the remarkable *Bored Sphere Problem* that first seems to have seen the light of day in Samuel I. Jones's *Mathematical Nuts* of 1932. A hole of length one unit is drilled through a sphere. What is the volume of the remaining shape?

The answer is independent of how big the sphere is—hard to believe until you realize that to drill a hole of length one metre through the Earth for example the radius would have to almost equal that of the entire planet for otherwise it would not come out the other side! Since the answer is, we are implicitly told, the same in all cases, we can turn the tables on the questioner and deduce the answer immediately by looking at the extreme situation where $R = 1/2$. This is the degenerate case where the radius of the hole is zero, and so the remaining volume is equal to the volume of the sphere, which is $\pi/6$. The general situation can be pictured as in Figure 8.5. The required volume is.

$$V = V_{\text{sphere}} - V_{\text{cylinder}} - 2V_{\text{cap}}.$$

We now simply apply our formulae for the value of each of these terms and use the relationships $r^2 - a^2 = a$ and $R - a = 1/2$ that

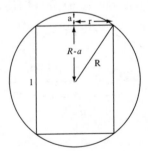

Fig. 8.5: The bored sphere

emerge in this particular set-up to simplify what we have. Miraculously all the terms in a and r cancel to leave you with the same answer every time: $\pi/6$

Volume of a Pyramid (pp. 79–80)

Let us look at the case of a triangular pyramid $ABCD$. We imagine it as part of a triangular prism also with the same base ABC and the same height h as shown in Figure 8.6.

The prism itself can be cut into three triangular pyramids: $ABCD$, $DEFB$, and $ABED$. Although these pyramids are *not* identical, any two of them have the same base and height, although which face is taken to be the base depends on what pair of them we are comparing. Our second pyramid has corners $DEFB$. If we removed it in order to take a good look at what remains we would see two triangular pyramids that we have separated in the Figure 8.7. Now the first two pyramids, $ABCD$ and $DEFB$ have equal bases, as $DEFB$ has as base the triangle EFD that is identical to the triangular base ABC of pyramid $ABCD$. They also have a common height h, that being equal to the height of the prism, and therefore they have the same volume. Now pyramids $DEFB$ and $ABED$ can be regarded as having a common base triangle, BED. What is more, looked at this way, each

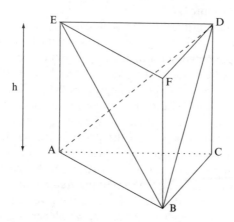

Fig. 8.6: Triangular prism formed from three pyramids

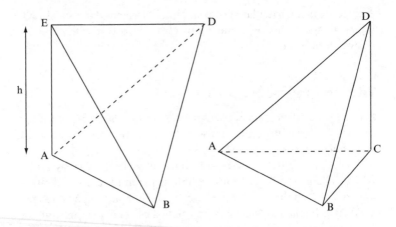

Fig. 8.7: Triangular pyramids separated

has the same height: the height of pyramid *DEFB* is the altitude of the triangle *EBF*, while the height of pyramid *ABED* is the altitude of triangle *EBA*; since these triangles are identical (they each form half the parallelogram *ABFE*) and since the altitude in each case is being measured from the common side *EB*, the two altitudes are the same. We conclude that these two pyramids also share the same volume and so all three do likewise. The volume of a triangular pyramid is therefore $\frac{1}{3}bh$.

Now let us consider the case of an arbitrary cone. The result that $V = \frac{1}{3}bh$ remains the same and can be reached by an argument based on approximating through triangles. Suppose the shape of the base is as pictured as in Figure 8.8 and that the apex of the cone is some point *P* at height *h* above the plane of the lamina.

Choose a point *Q* in the lamina as shown and approximate the perimeter of the figure with a many-sided polygon consisting of short lines between points on the perimeter. Consider the set of triangles formed by *Q* and these lines as indicated in the picture. Now the volume of the cone in question can be approximated as accurately as we please by the sum of the volumes of all the triangular pyramids with common apex *P* and the given triangular bases. Hence, using our formula for the volume of a triangular pyramid, we see that the volume of the cone is very nearly equal to $\frac{h}{3}$ times the

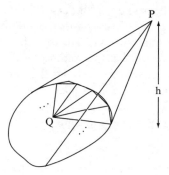

Fig. 8.8: Volume of a cone

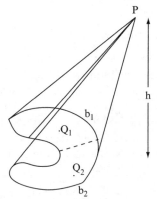

Fig. 8.9: Cone with arbitrary base

sum of the areas of all the triangles. The exact answer will be $\frac{h}{3}$ times the limiting value of this sum, which is evidently the area of the base of the lamina.

That is the reason why the general formula is a consequence of the result for triangular pyramids. The lamina of Figure 8.8 does not represent the most general case, however, as it has the special property of being *star-shaped*—we can choose a point Q which can be joined to every point on the perimeter curve without passing outside of the lamina.

The lamina in Figure 8.9 does not have that property. Nevertheless we can extend the argument to cope with this more general type of base. We break the base up, in this case into two star-shaped parts as shown, and the above discussion then demonstrates that the formula holds for the cone with apex P with each part in turn acting as base. In this way we see that the volumes of the smaller cones are $\frac{1}{3}b_1h$ and $\frac{1}{3}b_2h$ respectively say. Adding the two volumes together gives the volume of the full cone as $\frac{1}{3}bh$.

Riddle of the Age of Diophantus (p. 100)

The more literal translation goes like this:

God granted him to be a boy for the sixth part of his life, and adding a twelfth part to this, He clothed his cheeks with down; He lit him the light of wedlock after a seventh part, and five years after his marriage he granted him a son. Alas! late-born wretched child; after attaining the measure of half of his father's life, chill Fate took him. By consoling his grief by this science of numbers for four years he ended his life.

The paraphrased version in Chapter 4 p. 100 is perhaps easier to digest and recast as an equation by writing x for the length of Diophantus' life. The two sides of the equation correspond to the two ways in which we are told, in terms of x, the length of his son's life. On the one hand his son lived for $x/2$ years and on the other we also obtain an expression for the age of the son by subtracting from x all those parts of Diophantus' life that happened before his son was born or after his death. This leads to

$$\frac{x}{2} = x - \frac{x}{6} - \frac{x}{12} - \frac{x}{7} - 5 - 4.$$

We sort this equation out by taking all the terms in x across to the left, taking x out as a common factor, and finally doing the accompanying arithmetic, which leaves us with Diophantus enjoying an enviably long life of 84 years in all.

Although we know so little about him his influence has persisted through the ages. His main work was a treatise of thirteen books known as the *Arithmetica* and though only six of the books have survived into our own time, his name is remembered in the term *Diophantine equation* that applies to an equation for which the solutions are sought from numbers of a particular type, typically ordinary whole numbers or integers as they are called. The most famous example of such an equation is

$$x^n + y^n = z^n;$$

where solutions are required for the unknowns x, y, and z among positive integers. The solutions for $n = 2$ are called *Pythagorean triples* as they correspond to the sides of a right-angled triangle of whole number lengths—the equation then is no more than the statement of Pythagoras' Theorem for these triangles. All Pythagorean triples have been known since the time of Euclid. In the 1620s the great amateur French mathematician Pierre de Fermat scribbled in the margin of a Latin translation of the *Arithmetica* that the above Diophantine equation has no solution if n is more than 2 and that he had a marvellous proof to that effect only the margin was too small to accommodate it.[2] The first correct published solution to this problem, confirming that Fermat had been right all along, was produced by Andrew Wiles (now Sir Andrew) in 1994. It is hundreds of pages in length and Fermat's Theorem comes only as a consequence of many, many other things. This has quickly come to be regarded as the crowning achievement of 20th century mathematics. One of the more ancient and bizarre Diophantine equations was framed long before Diophantus' lifetime by Archimedes (287–212 BC). The notorious *Archimedes Cattle Problem* leads to a search for whole number solutions, x and y, to the equation

$$x^2 - (4\,729\,494)y^2 = 1.$$

The smallest solution was not found for over 2 000 years: in 1880 Amthor discovered it involves a number with 206 545 digits! This equation is an instance of what is known as *Pell's Equation*: a Diophantine equation of the form $x^2 - Ny^2 = 1$, where N is not itself a square. It was attributed to the 17th-century English mathematician Pell in error but the convenience of a one-syllable name has caused this misnomer to persist. It was studied much earlier, not only by the Greeks but by the 7th-century Indian mathematician Brahmagupta, as it arises in the process that the Greeks called *anthyphairesis* where one begins with two line segments and continues subtracting the shorter from the longer. This in turn is connected with the so-called *continued fraction representation* of $\sqrt{N}$ and to the golden

---

2. It seems that in this claim he may have been anticipated in the 12th century by
   Omar Khayyám, at least for the case $n = 3$.

ratio. A very readable reference on this topic is John Stillwell's *Mathematics and its History* (1991 and 2002).

Chapter 5

The Reflection Property of the Hyperbola (pp. 107–8)

The reflection property of the hyperbola is that a ray from one focus reflects as if it had come from the other and, as explained in the text, this is tantamount to verifying that the contact point P is the solution to the Heron Problem for the points F_2, G, and the tangent line L. Once again, the required picture is Figure 8.10, where, for brevity, we have renamed the foci as F and F' as shown.

We need to show that $GP + PF < GR + RF$ for any other point R on L. We first show that $GP + PF < GQ + QF$ where Q is the point of intersection of GR and the curve, although the argument applies to any point Q on this branch of the curve. Since G, P and F' are collinear we have

$$GP + PF' < GQ + QF'.$$

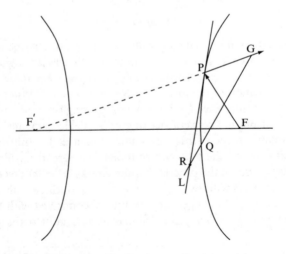

Fig. 8.10: Reflection property of a hyperbola

Adding PF to both sides we obtain:

$$GP + PF' + PF < GQ + QF' + PF$$
$$\Rightarrow GP + PF < GQ + QF' + (PF - PF').$$

Now, by our defining property for the hyperbola, the bracketed term is the same for *every* point on this branch of the curve, allowing us to replace $PF - PF'$ by $QF - QF'$ thereby giving:

$$GP + PF < GQ + QF' + (QF - QF') = GQ + QF$$
$$\Rightarrow GP + PF < GQ + QF.$$

The argument is now completed as in the case of the ellipse:

$$GP + PF < GQ + QF < GQ + (QR + RF) =$$
$$(GQ + QR) + RF = GR + RF.$$

Therefore P solves the Heron Problem and so we have established the hyperbolic reflection property.

The way in which we used an extension of the line $F'P$ to obtain our Heron interpretation throws further light on the parabolic case. If we consider the parabola to have a second focus 'at infinity' (to make proper sense of this requires the introduction of the common *direction* of the parallels to the axis to act as a new kind of point) then every conic has a pair of foci and this unifies our approach to the parabola and hyperbola, for in the case of the parabola we took an arbitrary point G to the right on the parallel through P to act as our second Heron point. In both the parabolic and hyperbolic cases G lies on the line from P on the conic to the second focus and is chosen so as to lie on the same side of the tangent line as the first focus, thereby allowing us to interpret the reflection problem as a Heron Problem. With these qualifications we can express the reflection property of all three conics in a single equivalent statement.

The contact point of a tangent to a conic is the solution of the Heron Problem for the tangent line and the foci.

Chapter 6

Nature of Isometries of the Plane (pp. 139–51)

Any isometry of the plane, that is to say, a mapping of the plane to itself that preserves distances, is either a translation, a rotation, a reflection, or a glide reflection. The first step in showing this is to demonstrate how any triangle $\triangle ABC$ can be sent by the four basic types of motions on to a given copy of itself, $\triangle A'B'C'$.

First, apply the translation that takes A on to A', giving us the situation as in Figure 8.11 where B has found itself translated to some point B_1. We can then move $A'B_1$ onto $A'B'$ by either a rotation around A' or, if we prefer, a reflection in the bisector of the angle $\angle B_1 A' B'$. In fact it can be shown that AB may be shifted to A'B' by either a direct or an indirect isometry.[3]

The effect of the chosen pair of isometries on the third vertex C is to shift it to some point C_1 and since isometries preserve distance either $C_1 = C'$ or C_1 is the reflection of C' in the line $A'B'$ (as a triangle is determined by the lengths of its sides: see pp. 45–6). In the

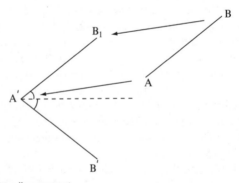

Fig. 8.11: Moving a line segment

---

3. If AB and $A'B'$ are parallel then one can be translated on to the other. Otherwise it is an interesting exercise to verify that AB can be *rotated* onto $A'B'$.

latter case, reflection of the triangle $\triangle A'B'C_1$ in the line $A'B'$ transforms the triangle on to $\triangle A'B'C'$. We conclude that a triangle may always be shifted to a congruent copy of itself using three or fewer of the basic isometries.

We have shown that there is an isometry α say, taking $\triangle ABC$ on to $\triangle A'B'C'$. Next we show that it is unique. Suppose that θ is any isometry that does this job and let D be any point in the plane of our triangles whose respective distances to the points $A, B,$ and C are $a, b,$ and $c,$ say. The point D is moved by θ to some point D' at respective distances $a, b,$ and c from $A', B',$ and C' and so lies on the intersection of the three circles with centres $A', B',$ and C' of radii $a, b,$ and c respectively. Now three different circles whose centres are not in line (as is the case here) can have at most one common point D', so that θ has only one choice as to where it maps D. We conclude that once an isometry α is constructed that maps a triangle on to a copy of itself then any isometry θ that does the same must behave the same as α not only on the triangle $\triangle ABC$ but on *every* point D in the plane, that is to say, $\theta = \alpha$.

This all serves to demonstrate that *every* isometry of the plane is a composition of three or fewer of the basic types of translation, rotation, reflection, and glide reflection. Indeed, it is not hard to show that a combination of two or more of the basic types simply yields another isometry of one of the four types. On our way to showing this, it is worth noting that any basic isometry can itself be built up of reflections alone. It is easy to check by examining the various cases that the effect of a pair of reflections in parallel lines separated by a distance d is that of a translation through a distance $2d$ perpendicular to the lines of reflection in the direction from the first line to the second. It follows that simulation of a translation through a distance d by reflections can be achieved using a pair of parallel lines separated by a distance $d/2$ at right angles to the direction of the translation. Similarly, the net effect of two consecutive reflections in lines with common point O that meet at an angle θ is to rotate all points about O through an angle 2θ in the direction from the first line to the second: again this is readily checked by looking at the various cases that arise and indeed the angles involved add in just the same fashion as do the distances in the case of parallel reflection axes. Conversely then, a rotation about a point O through an angle θ can be effected by a pair of reflections in lines through O that are separated by half that angle. Finally, since a glide reflection is the effect of a translation (that can be decomposed into two reflections) followed by another reflection, it follows that each of the basic types of isometries can be expressed as a combination of no more than three reflections and indeed, in the case of direct isometries, only two are ever needed. Furthermore, since any isometry is a combination of the basic ones, it follows that any isometry is expressible as the product of reflections: that is to say, it can all be done with mirrors.

In order to establish our claim that every isometry is one of the basic four (and not just a composition of them) we may use what we have just proved and demonstrate that the composition of any number of reflections can be written as one of the four types. Certainly this is true of one reflection so in order to finish it remains only to check that each of the four types followed by a reflection yields another isometry that is still a translation, a rotation, a reflection, or a glide reflection.

As we have just seen, the composition of two reflections is either a translation or a rotation depending on whether or not the lines of reflection are parallel.

Next, take a translation, followed by a reflection. If the direction of the translation and the axis of reflection are the same, then we already have a glide reflection. If not, let L be the line parallel to the reflection line L' that is taken by the translation on to L'. Take any two points A and B on L. Let M be the line parallel to both L and L' and halfway in between as shown in Figure 8.12 and take any point C on the line M. Let d be the distance the translation takes each point in the direction parallel to the line L'.

Consider the combined effect of our two isometries on the points A, B, and C. The points A and B are translated to points A' and B' on L' and are then left untouched by the reflection in L' while C is first translated to C_1, as shown, and then reflected to C' which is also on M. This trio of movements is also the effect of acting the glide reflection that shifts all points a distance d to the right and then reflects in the line M. By uniqueness of isometries it follows that the combined effect of the original translation and reflection is equal to the glide reflection just constructed.

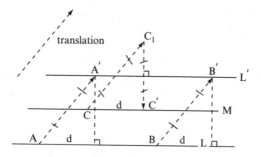

Fig. 8.12: Translation and reflection yield a glide reflection

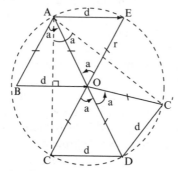

Fig. 8.13: Translation and rotation is a rotation

Next we look at the case of a glide reflection followed by a reflection. This can be regarded as a translation followed by a pair of reflections. In turn, the reflection pair is equal to either a translation or a rotation so we have either one translation followed by another, in which case the net effect is that of a translation whose vector (its associated directed line segment) is the sum of the two vectors of the translations in question, or we have a translation followed by a rotation that we shall now show yields another rotation through the same angle but with a different centre of turning.

Let O be the rotation centre and consider the isosceles triangle $\triangle AOE$ as shown in Figure 8.13 with base of length d, representing the vector of the translation, and whose angle a at O is the rotation angle; we let r stand for the radius of the circle centred at O through A and line segments of length r are marked in the diagram.

We also take the direction of *a* opposite to that of the translation so that the point *A* is fixed by the combined effect of the translation followed by the rotation. (If the translation was opposite the direction of the rotation we would work with the triangle that is the reflection of $\triangle AOE$ in the line parallel to *d* through *O*.) The effect of the translation and rotation is that of a rotation about the fixed point *A* through an angle *a*: to verify this we find three non-collinear points that are mapped by this rotation in the same way as they are under the translation–rotation pair. By construction, *A* is fixed in both cases. The point *B* is the point moved by the translation to *O* (which is then fixed by the original rotation) and *B* is also mapped to *O* by our rotation about *A* through angle *a*. Finally, the point *C*, the reflection of *A* in the line *BO*, is moved to *D* by the translation and on to *C'* by the rotation about *O* in such a way that $AC = AC'$; moreover $\angle CAC'$, being half the angle subtended at the centre of the circle, is also *a*, showing that *C* is moved to *C'* by the rotation about *A* through angle *a*, as required.

Our final problem is to describe as a single basic isometry the effect of a rotation about a point *O* through an angle *a* followed by a reflection in a line *L* as pictured in Figure 8.14. The point *O* is fixed by the rotation and then reflected through *L* to *O'*. The point *A* is chosen so that it is mapped by the rotation to $A' = B$, the intersection of *L* with the line *OO'* and, of course, *A'* is fixed by reflection in *L*. The point *B* is then rotated as shown to a point B_1 and then

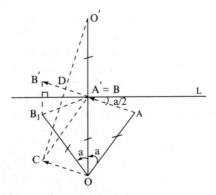

Fig. 8.14: Rotation and reflection gives glide reflection

reflected in L to B' and, by the symmetry of the diagram, A, B, and B' are collinear and the angle $\angle AA'L$ is $a/2$. We see then that the line segment AB is transformed by the rotation–reflection pair into the collinear line segment $A'B'$. Consider then the glide reflection with reflection axis and translation defined by the line segment AB. This glide reflection evidently maps A to A' and B to B' and to be able to conclude that it is the product of the original rotation–reflection pair it remains only to verify that the glide reflection also takes O to O'. Take OC parallel and equal to AB as shown and join C to O', meeting AB' at a point D. The claim is that O' is the reflection of C in the line AB. We observe that $BC = BO'$, as $ABCO$ is a parallelogram, and further we note the equality of angles

$$\angle O'BD = \angle OBA = (90 - \tfrac{a}{2})° = \angle OAA' = \angle CBD.$$

From this we infer that the triangles $\triangle CBD$ and $\triangle O'BD$ are congruent (SAS) and this gives the required facts that $CD = O'D$ and $\angle CDB = \angle O'DB = 90°$.

In the case where the reflection line L passes through O the result is the same but now the points O, A, and B coincide and we find that the product of the rotation about O followed by reflection in L is simply a reflection about the line through O inclined at an angle $a/2$ to L in the opposite sense to the rotation: that is to say, the translational element of the glide reflection vanishes.

Chapter 7

Geometry of the Pentagon (pp. 161–5)

The internal angle of a pentagon is $(2 \times 5 - 4)/5 = 6/5$ right angles which is $108°$. As $\triangle EAB$ in Figure 8.15 is isosceles one quickly sees that the angle between a diagonal and side such as $\angle AEB = 36°$ and then that the angle between two diagonals such as $\angle BEC = 108° - 2 \times 36° = 36°$ also. (This coincidence is explained by considering the circumscribing circle of the pentagon: angles $\angle AEB$ and $\angle BEC$ are equal as they stand on arcs of equal length.) We then infer that EP and DC are parallel as $\angle PED + \angle EDC = 72° + 108° = 180°$ and similarly PC is parallel to ED and so $DEPC$ is a parallelogram which is

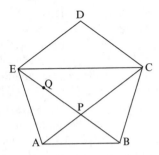

Fig. 8.15: Symmetries of the regular pentagon

indeed a rhombus as ED and DC are both of unit length and therefore so are its other two sides. Let d be the common length of the diagonals and consider the similar triangles $\triangle EPC$ and $\triangle APB$. Taking corresponding ratios reveals:

$$\frac{EP}{PB} = \frac{EC}{AB} = d = \frac{EB}{DC} = \frac{EB}{EP},$$

where the last equality uses the fact that DC and EP are opposite sides of a rhombus and are therefore equal. This says that the diagonal, EB, is divided by P in such a way that the ratio of the larger part, EP, to the smaller part, PB, is equal to the ratio of the diagonal EB to the larger part EP.

Since $EP = 1$ we see that $1/PB = d$ and so

$$EB = EP + PB \Rightarrow d = 1 + \frac{1}{d} \Rightarrow d^2 - d - 1 = 0,$$

and solving this reveals the value of the golden ratio d to be

$$d = \frac{1 + \sqrt{5}}{2}.$$

The self-generating nature of the golden ratio, d, comes about through the algebraic relationship $d + 1 = d^2$ or the equivalent $1/d = d - 1$.

To witness its self-propagating property mark the distance PB along PE to define the point Q as shown. From $1/d = d - 1$ comes $1/d^2 = 1 - 1/d$ which gives, since $PQ = PB = 1/d, EQ = 1 - 1/d$ and $PE = 1$

$$(PQ)^2 = EQ = (EQ)(PE),$$

which means that

$$\frac{PQ}{PE} = \frac{EQ}{PQ},$$

and so Q divides the smaller segment in the golden ratio: transferring QE on to the line QP would then give a point R that would divide QP in the golden ratio and we could continue *ad infinitum*. By a similar argument the process unfolds in the opposite direction: extending BE beyond E to a point R, such that $ER = PE = 1$, the claim is that

$$\frac{BR}{BE} = \frac{BE}{PE}$$

which, upon cross multiplying, amounts to the relationship $d + 1 = d^2$ that we have already seen as characteristic of the golden ratio, d.

Verifying that the length BP in Figure 7.10 is d then follows from applying Pythagoras to the triangle $\triangle MAD$, whereupon it follows that the golden ratio, and therefore the regular pentagon, is constructible.

Fermat Numbers and Factorization of $a^n \pm 1$ (pp. 164–5)

Numbers of the form $a^n \pm 1$ can rarely be prime. In the case of the minus sign we have the factorization:

$$a^n - 1 = (a - 1)(a^{n-1} + a^{n-2} + \cdots + a + 1);$$

and so we see that $a^n - 1$ cannot be prime unless $a = 2$. Even in this case, however, if n is a composite number, $n = ab$ say, then

$$2^n - 1 = (2^a - 1)(2^{a(b-1)} + 2^{a(b-2)} + 2^{a(b-3)} + \cdots + 2^a + 1);$$

and so $2^n - 1$ is itself composite. For example, for $n = 12 = 4 \times 3$ we have a factorization:

$$4\,095 = 2^{12} - 1 = (2^4 - 1)(2^8 + 2^4 + 1) = 15 \times 273.$$

The problem of primality of $a^n - 1$ is therefore reduced to the case of what are known as the *Mersenne numbers*, $2^p - 1$, where p is itself prime. Moreover the Mersenne numbers are not always prime, for example, $2^{11} - 1 = 2\,047 = 23 \times 89$. All the same it can be proved that any proper factor of a Mersenne number has the special form

$2kp + 1$ for some integer k.[4] The Mersenne primes are famously linked to another ancient aspect of number theory—the search for *perfect numbers*: numbers such as 6 and 28 that equal the sum of their factors apart from themselves; $6 = 1 + 2 + 3$, $28 = 1 + 2 + 4 + 7 + 14$. Euclid proved that *if* $2^p - 1$ *is prime* then the number $2^{p-1}(2^p - 1)$ is perfect. In the 18th century Euler proved the converse, in that every even perfect number arises in this way. (The existence of any odd perfect numbers remains a mystery to this day.) For example, the first four primes, 2, 3, 5, and 7, yield the first four Mersenne primes, 3, 7, 31, and 127 and so applying Euclid's result we obtain four even perfect numbers: 6, 28, 496, 8 128 and by Euler's converse implication we know there are no others in this range. Another snippet of information is that the even perfects alternately end in 6 and 28.

As regards the numbers of the form $a^n + 1$ with a at least 2, it is clear that if a is odd then $a^n + 1$ is even and so not prime. Moreover, if n itself has an odd factor m with $mt = n$, say, then $a^n + 1$ is composite as it is the subject of the following factorization:

$$a^n + 1 = (a^t + 1)(a^{(m-1)t} - a^{(m-2)t} + a^{(m-3)t} - \cdots + 1).$$

For instance, for $a = 2$ and $n = 9 = 3 \times 3$ we have:

$$513 = 2^9 + 1 = (2^3 + 1)(2^6 - 2^3 + 1) = 9 \times 57.$$

We do need m to be odd in this factorization in order that the alternating pattern of plus and minus signs finishes with $+1$, which is necessary to ensure that the expansion of the right-hand side telescopes down correctly, leaving just the two terms on the left. We conclude that a number of the form $a^n + 1$ is not prime unless a is even and n itself is a power of 2. For $a = 2$ we are left with only the Fermat numbers, $F_n = 2^{2^n} + 1$, as prime candidates.

Returning to the Fermat conjecture, let us look once more at the question of the primality of F_5. Since it exceeds 4 billion, verifying that F_5 is prime would involve division by each prime up to the square root of F_5 (as any composite number must have a prime factor that does not exceed its square root). Since

---

4. At the time of writing the largest known prime is the Mersenne number $2^{13,466,917} - 1$ that contains 4 053 946 digits: it would take practically a month just to write it down. Its primality was verified by the twenty year-old Canadian, Michael Cameron.

$$(2^{2^n})^2 = 2^{2^n + 2^n} = 2^{2 \times 2^n} = 2^{2^{n+1}},$$

we see that, for larger n, F_n^2 is approximately equal to F_{n+1}, as the 1 added is tiny compared to the powers of 2 involved. In other words, in order to prove that Fermat was right we would need to find all primes up to $F_4 = 65\,537$ and test each one for being a factor of F_5. It is no wonder that Fermat did not get around to doing this.[5] The hint that 641 might just bear a special relationship with F_5 comes by way of the observations:

$$641 = 2^4 + 5^4 = 5 \times 2^7 + 1.$$

To continue, the reader needs to have some familiarity with arithmetic modulo a prime p, that is, the arithmetic of the remainders left when numbers are divided by p. The number 641 is prime. (It is not too hard to check that it is divisible by no prime up to and including 23 and $p^2 > 641$ for any prime p over 23 so that is as far as we need go.) In particular we make use of the fact that arithmetic modulo the prime 641 is a field in which we may add, subtract, multiply, and, importantly here, divide (as long as we avoid division by numbers that are multiples of 641 as they leave 0 as remainder when divided by 641, and, as always, we cannot divide by 0.) Working modulo 641 the above pair of relations can be written as:

$$2^4 + 5^4 \equiv 0 \Rightarrow \frac{2^4}{5^4} + 1 \equiv 0 \Rightarrow \frac{2^4}{5^4} \equiv -1 \ (\text{mod } 641). \qquad (1)$$

While the second relationship can be expressed as:

$$5 \times 2^7 + 1 \equiv 0 \Rightarrow 5 \times 2^7 \equiv -1 \Rightarrow 2^7 \equiv -\frac{1}{5} \ (\text{mod } 641);$$

multiplying both sides of this by 2 we obtain:

$$2^8 \equiv -\frac{2}{5} \ (\text{mod } 641);$$

and raising both sides of this to the power 4 then yields:

$$2^{32} \equiv \left(-\frac{2}{5}\right)^4 \equiv \frac{2^4}{5^4} \ (\text{mod } 641);$$

5. The story is recounted in E. T. Bell's *Men of Mathematics* (1937) of the amazing calculator boy, Zerah Colburn, who, when asked if F_5 were prime, answered after a short delay that it was not, being divisible by 641. He could not explain however how he had worked it out!

but then our equation (1) allows us to say that

$$2^{32} \equiv -1 \;(\mathrm{mod}\; 641);$$

and by adding 1 to both sides we find what we seek:

$$F_5 = 2^{32} + 1 \equiv 0 \;(\mathrm{mod}\; 641),$$

that is to say, F_5 is not prime, it being a multiple of 641.

Fermat was an amateur mathematician, a judge by profession, and so he did not feel obliged to publish all his work, nor to prove everything that he discovered. In consequence his notes are full of unproved or partially proved assertions including the famous Last Theorem. Now that this has been resolved, we know that his conjecture on Fermat numbers was the one and only time he was wrong. Modern-day mathematicians are very fond of Fermat, who is often referred to as 'The Prince of Amateurs'. They feel for their hero and regret this blemish and for that reason are quick to point out that this was the only assertion of Fermat about which he expressed a degree of doubt.

Morley's Theorem on Angle Trisectors (p. 170–1)

The theorem can be proved by working in reverse; we begin with the equilateral triangle $\triangle PQR$ and build a general triangle that we identify with the given triangle $\triangle ABC$ after the fact. The adjacent angle trisectors meet to form $\triangle PQR$.

Although the meaning of the theorem is quite clear from the Figure 8.16, perhaps we should clarify the meaning of *adjacent* in this context. To each trisector T of an angle A of the triangle, associate the side of the triangle at the angle A that is closer to T than the other. Each side is thereby associated with two trisectors that are then described as (mutually) *adjacent*.

We will make use of a basic fact about the incentre, I, of a triangle $\triangle ABC$ (the intersection of the angle bisectors) that being that the $\angle BIC = 90° + A/2$ (see Fig. 8.17). Since I is the intersection of the bisectors we get immediately by considering $\angle BIC$ that,

$$\angle BIC = 180° - \frac{B}{2} - \frac{C}{2} = 180° - \frac{B+C}{2} = 180° - \frac{180° - A}{2} = 90° + \frac{A}{2}.$$

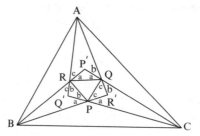

Fig. 8.16: Adjacent trisectors form an equilateral triangle

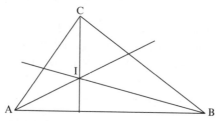

Fig. 8.17: Position of incentre

Indeed it follows that I is the *unique* point P lying on the bisector of the angle at A satisfying $\angle BPC = 90° + A/2$.

Returning now to our equilateral triangle $\triangle PQR$, on the sides QR, RP, and PQ, place isosceles triangles $\triangle P'QR, \triangle Q'RP$, and $\triangle R'PQ$, the base angles of which a, b, and c are chosen with some freedom but nevertheless satisfy:

$$a + b + c = 120°, \ a, b, c < 60°. \tag{2}$$

We extend the sides of the three isosceles triangles below their bases until they meet in the three points A, B, and C. Some angles can now be determined. For instance, since $\triangle RPQ$ is equilateral we have

$$\angle P'QA + a + c + 60° = 180°,$$

whence (2) gives: $\angle P'QA = 120° - (a + c) = b$. Similar justifications account for other angles in the diagram marked a, b, or c. The angle at A in $\triangle AQR$ is seen to be

$$
\begin{aligned}
180° - 2a - b - c &= 180° - (a + b + c) - a \\
&= 180° - 120° - a = 60° - a.
\end{aligned}
\tag{3}
$$

Furthermore, the extensions of the sides of the isosceles triangles do meet for, if $P'Q$ were parallel to $Q'P$, for instance, we would infer that

$$(a + 60°) + (60° + b) = 180°$$

which yields that $a + b = 60°$ whence we obtain $c = 120° - (a + b) = 120° - 60° = 60°$, contrary to our restriction on angles in (2).

Now $P'P$ bisects the angle at P' (since $P'RPQ$ is a kite with $P'P$ its axis of symmetry) and

$$\angle BPC = 180° - a = 90° + (90° - a) = 90° + \frac{1}{2} \text{ angle at } P'.$$

Since P lies on the bisector of the angle at P' it follows that P is the incentre of $\triangle P'BC$; in the same way Q and R are the incentres of $\triangle Q'CA$ and $\triangle R'AB$ respectively. In particular the incentre properties of P and Q respectively give that:

$$\angle BCP = \angle PCQ \text{ and } \angle PCQ = \angle QCA;$$

and so all three small angles at C are equal. Likewise this is true for the trio of small angles at A and at B.

By (3) the three small angles at A are each with value $\frac{1}{3}A = 60° - a$; similarly at B and at C and so we conclude that:

$$a = 60° - \frac{1}{3}A, \ b = 60° - \frac{1}{3}B, \ c = 60° - \frac{1}{3}C.$$

It follows that we have the freedom to accommodate any possible values for the angles A, B, and C with our choice of a, b, and c and so $\triangle ABC$ can be chosen to be similar to any given triangle. This proves that the adjacent trisectors of any given triangle form an equilateral triangle.

Impossibility of Duplicating the Cube and the Other Delian Problems (pp. 171–4)

The reciprocal of a number of the form $a + b\sqrt{c}, (b \neq 0)$, still has the same form: one can verify directly that

$$(a + b\sqrt{c})^{-1} = \left(\frac{a}{a^2 - cb^2}\right) + \left(\frac{-b}{a^2 - cb^2}\right)\sqrt{c};$$

and the expression itself is arrived at through the usual algebraic device of *rationalizing the denominator*, that is to say, multiplying top and bottom by the *conjugate surd* $a - b\sqrt{c}$. However there is a little more to be said. We are assuming that a, b, and c are each from some field of numbers F so that the bracketed coefficients are also members of F as a field is closed under the four ordinary arithmetical operations. What is more we are also assuming that $\sqrt{c}$ is *not* in F and that ensures that the denominator $a^2 - cb^2 \neq 0$, for if the denominator were 0 then we would find that $\sqrt{c} = \pm\frac{a}{b}$, which is not possible as this latter quantity lies in the field F.

In order to show that $x = \sqrt[3]{2}$ is not constructible the first task is to check that x is irrational and this can be done in much the same way as we show the same is true for $\sqrt{2}$ (see p. 57): we suppose that x equals some rational number $\frac{a}{b}$ cancelled to lowest terms, cube both sides, and then deduce a contradiction by showing first that a, and in consequence b also, is even. This shows that x does not lie in our basic field F of rationals but let us assume that it nonetheless is constructible from a unit length using Euclidean tools. We shall need to use the square root operation at least once on some rational, c say, in order to escape from F (which we henceforth call F_0), to a larger field, F_1, of all numbers of the form $a + b\sqrt{c}$, and perhaps we shall need to go beyond F_1 in the same fashion to fields F_2, F_3, etc. Let us denote by k the minimum possible number of new fields that we need to introduce in order to construct the cube root of 2, noting that $k \geq 1$, and so we may write

$$x = a + b\sqrt{c}, \text{ where } a, b, c \text{ lie in } F_{k-1} \text{ but } \sqrt{c} \text{ does not;}$$

where F_{k-1} is the field created previous to F_k.

The equation $x^3 = 2$ now becomes $(a + b\sqrt{c})^3 = 2$ and multiplying out this cube yields the equation:

$$p + q\sqrt{c} = 0, \text{ where } p = a^3 + 3ab^2c - 2, \text{ and } q = 3a^2b + b^3c.$$

Next we spot that $q = 0$, for if not we would get $\sqrt{c} = -\frac{p}{q}$, which is not possible as the right-hand side lies in F_{k-1} while the left does not. Therefore $q = 0$, which immediately forces p to be 0 as well.

The proof is finished by arriving at a contradiction that exploits the fact that roots of equations often come in conjugate pairs of the form $a \pm b\sqrt{c}$: the reader will no doubt be aware that this is the form for the roots provided by the quadratic formula and in general the study of the symmetries among the roots of polynomial equations was the motivation for the introduction of the theory of groups by Lagrange, Galois, and others in the late 18th and early 19th century. Here we need only note that if we put $y = a - b\sqrt{c}$ then $y^3 - 2 = p - q\sqrt{c}$, but since we have found that $p = q = 0$ we infer that y is also a cube root of 2. However, there is only one cube root of 2 (if you like, the graph of the function $f(x) = x^3$ cuts the horizontal line $y = 2$ just once) and so $x = a + b\sqrt{c} = y = a - b\sqrt{c}$; but this immediately tells us that $b = 0$ and so $x = a$—however this is once again an impossibility as a lies in the field of numbers F_{k-1} although x does not. This final contradiction now forces the conclusion that it was impossible to construct the cube root of 2 using Euclidean tools in the first place. The cube root of 2 is not a constructible number.

It is not difficult to show that the fields of numbers that arise when carrying out Euclidean operations all consist of *algebraic numbers*, that is to say, numbers that are solutions of polynomial equations with integer coefficients. In 1882 Lindermann showed that π was not a number of this kind but belonged to the class of so-called *transcendental* numbers which lay beyond the reach of such polynomial equations and so beyond the grasp of the Euclidean tools. It follows that $\sqrt{\pi}$ is also transcendental and therefore it is impossible to square the circle of unit radius because its area is exactly π.

Not all algebraic numbers are constructible: for example, $x = \sqrt[3]{2}$ is not constructible despite being algebraic (for it satisfies the polynomial equation $x^3 - 2 = 0$). In particular it can be shown that, from a given unit length, it is not possible to construct with straight edge and compasses a length that is a solution of a cubic equation with integer coefficients that lacks rational roots (and indeed $x^3 - 2 = 0$ is an instance of such an equation whose root, as we have seen, is not constructible). This is enough to show that it is generally impossible to trisect angles for we begin with the easily verified trigonometric identity:

$$\cos 3\theta = 4 \cos^3 \theta - 3 \cos \theta,$$

and put $\theta = 20°$. This yields that $x = \cos 20°$ satisfies the equation $8x^3 - 6x - 1 = 0$. This equation is cubic with integer coefficients yet has no rational roots. (By the Rational Root Theorem the only possibilities are ± 1, $\pm \frac{1}{2}$, $\pm \frac{1}{4}$, and $\pm \frac{1}{8}$ and none of them work.) It follows from the previously quoted theorem that the *length* $\cos 20°$ is not constructible, from which it readily follows that the *angle* $20°$ is likewise not constructible and so a $60°$ angle cannot be trisected using Euclidean tools even though a $60°$ angle is itself eminently constructible. Seeing as we cannot construct a $20°$ angle, the same goes for the regular 18-gon, as that polygon has an external angle of $20°$. If we could construct the regular nonagon, that is, the regular nine-sided polygon, we could, by bisecting its sides, build the 18-gon. Since this is impossible, it follows that the regular nonagon also cannot be constructed by using straight edge and compasses alone.

Further Reading

Books on mathematics and its rich history are published frequently enough and can be found in any serious bookshop. Older books are often real treasures and you are most likely to come across them at a library. The production of these older texts is less slick so they may appear uninviting and the writing style is generally more terse and less conversational but if you can see your way past this you will find these books are often better written, can be wonderfully clear, and are mathematically action-packed. Although style may date, mathematical content hardly does at all and so you may find some of the older books surprisingly rewarding to read and refreshingly free from distractions.

There are a number of large-scale books dedicated to the history of mathematics suitable for serious study. Among these are *A History of Mathematics*, by Carl B. Boyer (New York: Wiley, 1968) and *An Introduction to the History of Mathematics* by Howard Eves (New York: Holt, Rinehart and Winston, 1969). For a more biographical approach, E. T. Bell's *Men of Mathematics*, some of whom are women (New York: Simon and Schuster, 1937) is always popular. The BBC book of the series *The Ascent of Man* by J. Bronowski (London: BBC, 1976) and Bertrand Russell's classic *History of Western Philosophy* (London: Allen and Unwin, 1961) are both written by mathematicians who appreciate and can explain the role of mathematics in history. Russell writes with such clarity and authority especially concerning Greek philosophy that the reader feels compelled to accept everything he says although philosopher colleagues assure me that there are alternative views. Russell and Bronowski's understanding of mathematics however is certainly sound. If you dare to grapple with what Russell really thought about mathematics there is his *Introduction to Mathematical Philosophy* (New York: Allen and Unwin). John Stillwell's *Mathematics and its History* (New York: Springer-Verlag, 1991 and 2002) is an unusual and excellent book that teaches general mathematics in a unified manner in the context of its historical development and it particularly features a lot of serious geometry. The snippet on the role of matrices in quantum physics featured in this book was taken from the account given in one of John Gribben's books *In search of Schrodinger's Cat* (London: Black Swan, 1994).

Recent popular books about things mathematical must include Simon Singh's *The Code Book* (London: Fourth Estate, 1999); he came to notice through his acclaimed *Fermat's Last Theorem* (London: Fourth Estate, 1997). There is a book with the same title by Amir D. Aczel subtitled *Unlocking the Secret of an Ancient Mathematical Problem* (London: Penguin, 1997). This is a short account, without Singh's many interesting digressions, that offers description of some unedifying rivalries and recriminations concerning priority among some of the main players—it is not all comfortable reading. *Fermat's Last Theorem for Amateurs* by Paulo Ribenhoim (New York: Springer-Verlag, 1999) is a mathematics book and a good one. Don't be taken in by the name, it is written for readers with a mathematical training who are not experts in number theory so they can capture some of the flavour of the mathematics surrounding the problem. It does end with an outline of the work of Faltings, Frey, Ribet, and Wiles that led to the eventual solution. Another excellent recent book for the serious mathematician is *Proofs from the Book*, by M. Aigner and G. Ziegler (London: Springer-Verlag, 1999). This volume, dedicated to the memory of Paul Erdos (whose fantasy was of a heavenly tome in which were written all the best proofs), consists of thirty extraordinarily beautiful yet relatively elementary proofs of some quite deep results such as Bertrand's Postulate that a prime always lies between n and $2n$, Wedderburn's Theorem that every finite division ring is a field, Hilbert's Third Problem on decomposing polyhedra, and the five-colouring of plane graphs to give just a sample. Short recapitulations of the relevant theory make the book very instructive and satisfyingly self-contained. Of course there are many good books about number theory but I have cited two, *An Introduction to the Theory of Numbers* by G. H. Hardy and G. M. Wright (Oxford: OUP). First published in 1938 it is regularly updated and goes on selling. It is written in a masterly fashion, from the beginning it wastes no time and is brim full of action right up to Theorem 460 some 400 pages later. *The Book of Numbers* by John Conway and Richard Guy (New

York: Springer-Verlag, 1996) is full of history, vivid pictures, and all manner of facts about numbers. Not really a textbook, it is similar in one way to Hardy and Wright in that the authors are keen to explain everything interesting about numbers that they can.

Mathematicians continue to write books whose purpose is to combine some overall view of modern mathematics with as much real detail as is possible to convey to a general audience and an example is my own earlier book, *Mathematics for the Curious* (Oxford: OUP, 1998). Ian Stewart is a most energetic author in every way who has written a number of popular books along these lines such as *From Here to Infinity* (Oxford: OUP, 1996). If you can find a copy of his original *Concepts of Mathematics* (London: Pelican, 1975) it could be well worth your while. A general reader might find it very tough going and a bit too much like a mathematics text to be really enjoyable but for someone who already knows a little about higher mathematics it would be hard to find a better book that introduces and explains so much. The impossibility of Duplicating the Cube given in this book follows the very concise argument given there. At the other extreme one might say we have *Introducing Mathematics* by Z. Sardar, J. Ravetz and B. Van Loon (Cambridge: Totem Books, 1999). This cartoon version of mathematics is full of jokes but all the same is not shy of introducing some very deep ideas. Another viewpoint is provided by John Allen Paulos in *A Mathematician Reads the Newspaper* (New York: Basic Books, 1995). There are mathematical novels about: *The Parrot's Theorem* (London: Orion Fiction, 2000) by Denis Guedj, is a mystery devoted to Fermat's Last Theorem and the Goldbach Conjecture (every even number is the sum of two primes): quite extraordinary, as is *The Wild Numbers* by Philibert Schogt (London: Orion Fiction, 2000) which captures the feeling of research mathematics and the accompanying triumph and delusion in a way most readers will find surprising.

Two early proselytizing works are *Mathematics for the General Reader* by E. C. Titchmarsh (London: Hutchison Press, 1959) and *Mathematics for the Million* by Lancelot Hogben (London: Allen and Unwin, 1937). The first is a short, sober, and altogether dignified introduction to mathematics that is a pleasure to read. The second is an extraordinary and enthusiastic account, starting very much from scratch, full of idiosyncratic ideas interspersed with the occasional denunciation of Hitler. Chapter 8 gives a surprisingly detailed account of Spherical Geometry and astronomical applications are one of the features of the book. There are of course many other popular mathematics books written during the middle part of the previous century whose purpose was to explain the changing nature of the subject: the author W. W. Sawyer wrote a number and perhaps in deference to the title I should mention *Mathematics and the Imagination* by E. Kasner and J. Newman (London: Pelican, 1968): again quite readable with some interesting gems, such as an incredible construction, due to Brouwer, of a map of three countries in which every single point along the boundary of each country is a meeting place of all three countries! The little book by Mark Kac and Stanislaw Ulam, *Mathematics and Logic* (London: Pelican, 1971) is self-contained but is not shy of detail and is really quite ambitious in its scope. George Polya, whose work so influenced Escher, wrote a number of books that focus on the psychology of mathematics and how one can successfully go about doing it.

If you like puzzle books, it is hard to go past the 19th-century American collector Sam Loyd: *More Mathematical Puzzles* (New York: Dover Publications, 1960). The Master of the Puzzle Book was the late Martin Gardner, *Mathematical Puzzles and Diversions* (London: Pelican Books, 1961) and of course there are dozens of other books along these lines. However, *The Mathematical Olympiad Handbook* by Anthony Gardiner (New York: OUP, 1997) is well worth a look. It consists mainly of a compilation of the elementary though very tough questions used to test the talent in the Mathematical Olympiad Competition but does also offer a very useful and concise primer on the mathematical techniques involved in the problems, the likes of which would be hard to find elsewhere.

As regards astronomy there are many popular accounts but I drew upon the book by Norman Davidson *Astronomy and the Imagination: A New Approach to Man's Experience of the Stars* (London: Routledge & Kegan, 1985) as it takes the curious viewpoint of rekindling the idea that we should try, at least some of the time, to view the heavens exactly as they appear. Another good source are the books of I. B. Cohen among which are *The Birth of the New Physics* (London: Heinemann, 1961). Dava Sobel's bestseller *Longitude* (London: Fourth Estate, 1998) tells the story of the age-old problem and how Harrison and his time pieces eventually won the day despite the best efforts of some (but by no means all) in the scientific establishment to deny him his prize.

As for geometry, it is not easy to find copies of Euclid's *Elements* any more outside of well-stocked libraries. Older school textbooks such as *A School Geometry, I, II, and III* by H. S. Hall and F. H. Stevens are a good substitute. The same authors also wrote *A Textbook of Euclid's Elements* which gives a faithful rendition of some of the Euclidean books. These school books were first published in 1889 but continued in print at least till 1949 and so copies are still quite common, usually decorated with school-boy annotations. They are perfectly clear and readable books with no nonsense. However the introduction they give to Book 5, Eudoxes' *Theory of Proportion*, has the peculiar disclaimer that it is 'very rarely read' and advises the student to bypass it and to perhaps return later 'if it is thought desirable'. Of course, there are many geometry books but two good ones that I have made use of here are H. S. M. Coxeter's *Introduction to Geometry* (New York: Wiley, 1969). This is a textbook designed for a student taking three years of geometry so begins with general geometry, has mathematical preliminaries in the middle of the book, and deals with differential geometry and the like that requires calculus and complex numbers in the latter part. It is written by a true expert though and that is evident throughout. A second classical work is *Geometry and the Imagination* by David Hilbert and S. Cohn Vossen (New York: Chelsea Publishers, 1952). First published in 1932 this is a wonderfully readable book with excellent photographs and diagrams. Again, it is a mathematics book but is more exploratory and does not have the standard, Lemma–Theorem–Proof–Corollary–Remark structure of a textbook but all the same explains the mathematics properly. Being more discursive than a formal textbook, the level of difficulty varies throughout the book as required by the authors to explain what they think important. A modern problem book is *Unsolved Problems in Geometry*, by H. T. Croft, K. J. Falconer, and R. K. Guy (New York: Springer-Verlag, 1991). As the title suggests, it consists of outstanding geometric problems along with some explanation as to where the eventual solutions could lie. All these problems are of course extraordinarily difficult but some, such as that of finding the minimal perimeter of comfortable living quarters for the unit worm (so he can stretch out into any position he likes) can be explained to anyone. Another more recent effort is David Gays' *Geometry by Discovery* (New York: Wiley, 1998).

The Penrose tilings feature but are no means the central feature of Roger Penrose's speculative book on the nature of physics, *The Emperor's New Mind* (Oxford: OUP, 1989), while the standard textbook on tilings is *Tilings and Patterns* by B. Grunbaum and G. C. Shepard (New York: W. H. Freeman, 1987). Majorie Senechal's book *Quasicrystals and Geometry* (Cambridge: Cambridge University Press, 1995) is a unique mathematical introduction to the topic that is none the less accessible in part to a reader without a specialist mathematical background. Donald Hill provides an account of Islamic mathematics in *Islamic Science and Engineering* (Edinburgh: Edinburgh University Press, 1993). My account here of the 17 tiling groups is taken from the paper of Doris Schnattschneider in the *American Mathematical Monthly*, 85 (1978), 439–50. However, a complete derivation of the 17 plane crystallographic groups is in the recent book *Symmetries* by D. L. Johnson (London: Springer-Verlag, 2001). A delightful account of the pictures of Escher with straightforward explanation by the artist himself is provided in *M. C. Escher, The Graphic Work* (London: Taschen, 1992).

Acknowledgements

Chapter 6

Figures showing tiling patterns in this chapter are after Doris Schnattschneider, *American Mathematical Monthly*, 85 (1978), 439–50.

Chapter 7

Figures 7.17 (a) and (b) are after Marjorie Senechal, *Quasicrystals and Geometry*, (CUP, 1995).

Figures 7.18 and 7.19 are after (Sir) Roger Penrose, *The Emperor's New Mind*, (OUP, 1989).

Index